Perspectives in
Quantum Theory

The MIT Press
Cambridge, Massachusetts, and
London, England

Perspectives in Quantum Theory

Essays in Honor of
Alfred Landé

edited by
Wolfgang Yourgrau
and
Alwyn van der Merwe

Set in Monotype Baskerville
by Wolf Composition Co., Inc.
Printed by Murray Printing Co.
and bound by The Colonial Press Inc.
in the United States of America.

ISBN 0 262 24014 9 (hardcover)

Library of Congress catalog card
number: 73-123253

Alfred Landé

Introduction

Alfred Landé
and the Development
of Quantum Theory

Half a century is a very long time in an era that has seen the dramatic upheavals of modern physics, and few are its key actors that have thus long held center or side stage while not losing the faculty to create an intellectual stir. Alfred Landé, who last month celebrated his eighty-second birthday in the American city of Columbus, Ohio—his home since leaving his native Germany in 1931—is such a phenomenon. His creative scholarship came to the fore during the years of the First World War and flourished in the 1920s, coincident with the unparalleled and feverish quest of that decade for a breakthrough in man's understanding of the atom. But, as one look at Landé's bibliography (p. 270 ff.) would confirm, his researches and flow of ideas have continued beyond this youthful period, though in ways less conspicuous and increasingly on the philosophical side, to this very day.

The present book, a compilation of essays written by some well-known physicists and philosopher-scientists, represents a timely tribute by these authors and ourselves to Alfred Landé, both as a creator of undisputed new knowledge and as an unruffled, independent thinker in matters where general agreement is hard or impossible to come by. Most physicists will no doubt see Landé's claim to lasting fame as residing in the order he once established, with a brilliantly conceived quantitative hypothesis, in what was until then a jumble of confused spectroscopic data. But Landé himself believes that over the past decades we have been the victims of an obfuscation more serious because this time it involves our interpretation of the most fundamental concepts of the physical world. If in this supposition Landé is correct and if his vigorously advocated proposals again should furnish the simplifying key, then our present tribute would be doubly deserved.

The historic hypothesis alluded to in the foregoing paragraph is, of course, the famous *Landé g* or *gyromagnetic factor* (actually it is the inverse of the gyromagnetic factor), which determines the fine, as well as the hyperfine, structures of optical and x-ray spectra. Textbook writers usually exhibit this factor in a context that makes its scope appear unduly restrictive. To remove this defect and to clarify our nomenclature for later use, we may be permitted here a brief introduction to the g-factor from a (relatively) modern standpoint. Shunning the complicated apparatus of quantum mechanics proper, we invoke here the so-called vector model of the atom, which conceptual tool, fittingly, earned its accreditation initially through Landé's work. This approach keeps our derivation from passing the test of absolute rigor, but this shortcoming is more than compensated for by the visual gain and convenience of the vector model, if account is also taken of the fact that quantum mechanics would not change our final result in any way.

The decisive breakthrough to a better understanding of the anomalous Zeeman Effect came when Landé had the idea of the "term analysis" of the spectral Zeeman types in two magnetic energy patterns, each of them turning out to have its peculiar gyromagnetic factor g. It was the real Rosetta stone deciphering the hieroglyphics of the magnetic splitting effect. Previous investigators had always studied the visible Zeeman *types* with their complicated numerical regularities that now were reduced to the simple regularities of the magnetic energy *terms*. What came later, including the g formula (see below), was then a matter of routine to establish.

Consider then that part of the angular momentum of an electron, $\mathbf{L}$, which stems solely from its orbital motion. With it, we must obviously associate a magnetic dipole moment $\boldsymbol{\mu}$ that is antiparallel to $\mathbf{L}$; the spatial opposition of these two vectors being due to the negative value of the electronic charge $-e$. If $\mathbf{L}$ is represented by a quantum vector of magnitude $l^*\hbar$ and μ by a vector of magnitude $\mu_B\mu^*$, where $\mu_B = e\hbar/2m_0c = 9.273 \times 10^{-21}$ erg gauss^{-1} is the

Bohr magneton (m_0 denotes the electron mass), then the g-factor for the orbital motion is defined by

$$\mu_l^* = g_l l^*.$$

Similarly, given that the spin angular momentum **S** of the electron is a vector of length $s^*\hbar$, the g-factor of spin origin is related to the electron's intrinsic magnetic moment through

$$\mu_s^* = g_s s^*.$$

Thus, in both cases, the g-factor is nothing but the ratio of an elementary magnetic moment to its causative angular momentum, provided the latter quantity is measured in units of $\hbar$ (a natural choice since Bohr first postulated the quantization of orbital momentum) and the former in units μ_B.

An inspection of the Hamiltonian function of an electron in an external magnetic field reveals that $g_l = 1$ for any kind of electronic orbit (which explains why we choose to express magnetic moments in Bohr magnetons); whereas Dirac's relativistic theory of the electron requires that $g_s = 2$. (Quantum electrodynamics in the formulation of Schwinger and Tomonaga predicts, in more precise agreement with experiment, a slightly larger value, viz. $g_s = 2.00232$, but this refinement need not concern us.) Furthermore, quantum mechanics demands that we set the squared magnitudes of the vectors **L** and **S** (in units of $\hbar^2$) equal to $l^{*2} = l(l + 1)$ and $s^{*2} = s(s + 1) = \frac{3}{4}$, respectively, in terms of the orbital and spin quantum numbers l and s; so that the defining equations of g_l and g_s read, more explicitly,

$$\mu_l^* = g_l\sqrt{l(l + 1)} \qquad \text{and} \qquad \mu_s^* = g_s\sqrt{s(s + 1)}.$$

Complications arise when one considers a "mixed" vector **J** compounded from both orbital and spin momentum contributions. Then, due to the differing values of g_l and g_s, $\boldsymbol{\mu}$ is no longer anti-parallel to the momentum vector **J**. However, if $\mathbf{J} = \sum (\mathbf{L}_i + \mathbf{S}_i)$, the sum of the orbital and spin momenta of all the electrons in an isolated atom, then **J** (the total angular momentum of the electronic system) is constant in direction and magnitude; and, consequently, all the individual momenta $\mathbf{L}_i$ and $\mathbf{S}_i$, and thus also the magnetic moment $\boldsymbol{\mu}$ associated with $\mathbf{J}$, must precess about $\mathbf{J}$. For times long enough compared with the period of this precession (as will normally be the times of interest in spectroscopy), the component of $\boldsymbol{\mu}$ perpendicular to $\mathbf{J}$ will average to zero, so that only the component μ_J of the magnetic moment along $\mathbf{J}$ remains experimentally effective. Accordingly, we define the g-factor in this case by

$$\mu_J^* = gJ^* = g\sqrt{J(J+1)},$$

where the asterisks again denote measurements of magnetic moment in Bohr magnetons and angular momentum in units of $\hbar$, and where the capital letter J (j for a single electron) is the total angular-momentum quantum number determining the quantum-mechanical squared magnitude $J^{*2} = J(J+1)$ of the vector $\mathbf{J}$.

How does the general quantity g depend on the vectors $\mathbf{L}_i$ and $\mathbf{S}_i$? Obviously, addition of the magnetic-moment components along $\mathbf{J}$ gives

$$gJ^* = \sum_i \langle l_i^* \cos(\mathbf{L}_i, \mathbf{J}) + 2s_i^* \cos(\mathbf{S}_i, \mathbf{J})\rangle,$$

where the angular brackets designate a time average, and the rest of the notation is self-explanatory. On combining this with the addition formula

$$J^* = \sum_i \langle l_i^* \cos(\mathbf{L}_i, \mathbf{J}) + s_i^* \cos(\mathbf{S}_i, \mathbf{J})\rangle,$$

one gets

$$g = 1 + \sum_i (s_i^*/J^*)\langle\cos(\mathbf{S}_i, \mathbf{J})\rangle = 2 - \sum_i (l_i^*/J^*)\langle\cos(\mathbf{L}_i, \mathbf{J})\rangle.$$

This furnishes a general formula for the g-factor in two alternative forms. However, the result is of no use in practice unless the angles $(\mathbf{S}_i, \mathbf{J})$ or $(\mathbf{L}_i, \mathbf{J})$ are known. And, except when $\mathbf{J}$ is composed of only two vectors, such knowledge is available for not more than a few special situations. The usual practice of textbooks to confine analysis to one or two of these should, however, not obscure the fact that a wider formula, covering an infinite variety of possibilities, can be written down in the above manner.

The simplest type of situation giving rise to a practical formula occurs when what is known as *Russell-Saunders coupling* of momentum vectors obtains. In this instance it is assumed that the magnetic interaction among the $\mathbf{L}_i$, and also among the $\mathbf{S}_i$, is so strong that the elementary vectors first combine to give the resultant orbital vector $\mathbf{L}$ and spin vector $\mathbf{S}$, respectively, which then combine to form the total angular momentum $\mathbf{J} = \mathbf{L} + \mathbf{S}$ for the atom. The magnitudes squared $L^{*2} = L(L+1)$ and $S^{*2} = S(S+1)$ of $\mathbf{L}$ and $\mathbf{S}$ are known once the orbital quantum number L (l for one electron) and the spin quantum number S (s for one electron) are specified. This (extreme) state of affairs actually holds for at least some states of a large number of light elements in particular. All the $\mathbf{L}_i$ precess under these conditions about their resultant $\mathbf{L}$ (at a rate that increases with the strength of their interaction), which itself precesses about $\mathbf{J}$ (at a lesser rate); and the same is true for the $\mathbf{S}_i$ and their resultant $\mathbf{S}$. A short geometric consideration shows that $\langle\cos(\mathbf{S}_i, \mathbf{J})\rangle = \langle\cos(\mathbf{S}_i, \mathbf{S})\rangle \cos(\mathbf{S}, \mathbf{J})$. One has, moreover, $S^* = \sum s_i^* \langle\cos(\mathbf{S}_i, \mathbf{S})\rangle$. Substitution of these results, together with similar ones for the $\mathbf{L}_i$, in the above formulas for g, leads to

$$g = 1 + (S^*/J^*)\cos(\mathbf{S}, \mathbf{J}) = 2 - (L^*/J^*)\cos(\mathbf{L}, \mathbf{J}).$$

Finally, we may appeal to the cosine formulas, such as

$$2S^*J^*\cos(\mathbf{S}, \mathbf{J}) = J^{*2} + S^{*2} - L^{*2},$$

and insert the quantum-mechanical values of L^*, S^*, and J^*, to obtain

$$g = 1 + \frac{J(J+1) + S(S+1) - L(L+1)}{2J(J+1)},$$

the oft-quoted *Landé g formula.*

According to the quantum-mechanical addition law for angular momenta, the quantum number J can assume only the values

$J = L + S, L + S - 1, \ldots, |L - S|$, so that either $2S + 1$ (if $L \geq S$) or $2L + 1$ (if $L \leq S$) quantum states (LSJ) occur for specified values of the numbers L and S. These states, together forming a *term*, of *multiplicity* $2S + 1$, are split by the *spin-orbit* interaction (the potential energy of the spin magnetic moments in the magnetic fields generated by the orbiting electrons) into a compact group of energy levels distinguished by different values of J. If $L \geq S$, the number of *components* thus obtained equals the multiplicity of the term, causing us to speak (in any event) of *singlet*, *doublet*, *triplet*, etc., terms. The individual components or (sub)levels are given the designation ${}^{2S+1}X_J$, wherein X stands for one of the letters S, P, D, etc., corresponding to $L = 0, 1, 2$, etc.

The sublevel structure of the terms has its counterpart in the *fine structure* of the observable atomic spectrum: A close group of lines, or a *spectral multiplet*, appears as a result of quantum transitions from one term to another, insofar as these transitions are allowed by the selection rules for J, L, and S. Thus, as a simple example, the yellow sodium D_1 and D_2 lines (5896 Å and 5890 Å) are the components of a doublet, corresponding to the transitions $3^2P_{\frac{1}{2}} \rightarrow 3^2S_{\frac{1}{2}}$ and $3^2P_{\frac{3}{2}} \rightarrow 3^2S_{\frac{1}{2}}$, respectively, of electrons with principal quantum number $n = 3$.

The fine-structure splitting of a spectral line is, of course, dependent on the separations between the sublevels of the respective terms involved in its emission. These separations, which increase with the strength of the spin-orbit interaction, can be found, to within a factor, by the following vector-model argument. To a good approximation, one may neglect the interaction between the spin of an electron and the orbital magnetic moment of another electron, and keep only the interaction energy between $\mathbf{L}_i$ and $\mathbf{S}_i$ of the same electron. Semiclassical considerations suggest that the value of the latter is of the form $a_i\langle\mathbf{L}_i \cdot \mathbf{S}_i\rangle$, a result that is rigorously confirmed for a hydrogenlike atom by Dirac's relativistic theory of the electron. The spin-orbit energy thus becomes approximately $E_{so} = \sum a_i l_i^* s_i^* \langle\cos(\mathbf{L}_i, \mathbf{S}_i)\rangle$. In the Russell-Saunders coupling scheme, we can write this as

$$E_{so} = \cos(\mathbf{L}, \mathbf{S}) \sum_i a_i l_i^* s_i^* \langle \cos(\mathbf{L}_i, \mathbf{L}) \rangle \langle \cos(\mathbf{S}_i, \mathbf{S}) \rangle$$

$$= AL^*S^* \cos(\mathbf{L}, \mathbf{S});$$

or, since $2L^*S^* \cos(\mathbf{L}, \mathbf{S}) = J^{*2} - L^{*2} - S^{*2}$,

$$E_{so} = (A/2)[J(J+1) - L(L+1) - S(S+1)],$$

with A depending on L and S but not on J. As a consequence, the energy interval between the adjacent levels J and $J - 1$ of a term becomes $\Delta E = (A/2)[J(J+1) - (J-1)J] = AJ$, that is, proportional to the J value of the upper level. This result is the *Landé interval rule*, which is fairly well obeyed in all cases of Russell-Saunders coupling and which greatly facilitates the empirical analysis of spectral multiplets.

The alternative designation of g as the *splitting factor* arises from the direct explanation it affords of the spectral behavior of atoms in an external magnetic field. Under normal conditions, the energy levels of an atom are defined by the configuration of its electrons and the quantum numbers L, S, and J (where we assume Russell-Saunders coupling and ignore hyperfine structure due to nuclear interactions or Lamb shifts). The quantum states of the atom, however, are not fixed unless one knows additionally the magnetic quantum number M_J, which is capable of assuming the values $J, J - 1, \ldots, -J$ and measures the component $M_J \hbar$ of the total angular momentum $\mathbf{J}$ along any physically distinguished direction. Thus $2J + 1$ states coincide in the level J; but we can remove this degeneracy by imposing a magnetic field $\mathbf{H}$ on the atom, since each of the $2J + 1$ possible orientations of $\mathbf{J}$ relative to this field direction will correspond to a different value of the magnetic energy $-\boldsymbol{\mu}_J \cdot \mathbf{H} = gJ^* \mu_B H \cos(\mathbf{J}, \mathbf{H}) = M_J g \mu_B H$.

The description according to the vector model of the magnetic splitting of a level J is similar to that of the spin-orbit resolution of a term with given values of L and S. Both rest on a theorem of classical electrodynamics (Larmor's theorem, 1897), according to which any magnetic moment acted on by a magnetic field will (in first approximation) precess around the field direction with a frequency that is proportional to the field strength. In the spin-orbit phenomenon, the magnetic moment associated with **S** precesses in the magnetic field attributable to **L**; and since this *internal* field is rather strong, the precession is rapid, and a relatively large level separation $-\boldsymbol{\mu}_S \cdot \langle \mathbf{H}_{\text{int}} \rangle$ characterizes the fine structure of the term in question. Likewise, in the magnetic splitting effect, the whole electronic system, and thus **J** and $\boldsymbol{\mu}_J$, precess about the direction of the external field **H**. However, as this field is usually weak compared with the internal field, the associated precession rate about **H** is smaller, and the separation of the $2J + 1$ magnetic levels is much less than the separation of the levels of the unperturbed term.

The level-splitting formula $\Delta E = M_J g \mu_B H$ explains both the *normal Zeeman effect*, consisting of the splitting of a spectral line into three components in a magnetic field, and the *anomalous Zeeman effect*, where the field causes the resolution of a line into more than three components. The deciding factor between these two effects is the multiplicity of the levels involved in the electronic transition. The normal effect can be observed on spectral lines arising from transitions between singlet states. For such states, $S = 0$ and thus (from Landé's g formula) $g = 1$; so that both the upper and lower energy levels of the transition will, when acted upon by a magnetic field, split into components with the same energy interval $\mu_B H$ between any two adjacent components. In conjunction with the selection rules $\Delta M_J = 0, \pm 1$, this fact leads to the conclusion that, no matter how numerous the magnetic components, only three different energy intervals, and thus only three separate spectral lines (symmetrically situated about the original line) are allowed to occur in transitions. The splitting into a triplet of the Cd 6438 Å line ($^1D \rightarrow {}^1P$ transition) is an illustration of this, rather infrequent, state of affairs. The anomalous Zeeman effect, which (despite its name) really is the more common phenomenon, occurs in all other cases, i.e. when one or both states of the transition has a multiplicity greater than unity (an S value different from zero). For then g, and thus the energy interval $g\mu_B H$, is different for the upper and lower states. Consequently, the line of the spectral multiplet under consideration will resolve into more than three components under the influence of an applied magnetic field. A familiar illustration of this is the earlier mentioned D_1 and D_2 lines of the Na D doublet, which split into four and six magnetic components, respectively.

In an external magnetic field that is large compared with the internal field of the atom, the Zeeman-level splitting predominates over the spin-orbit splitting. The result is a distortion of the Zeeman line patterns known as the *Paschen-Back effect*, in whose analytical description the g-factor no longer appears.

The formulas and ideas required in the analysis of atomic spectra under ordinary conditions, which were set out on the foregoing pages, are straightforward enough and certainly not difficult to comprehend. The amazing fact, though, is that Landé arrived semi-empirically at the g formula and related results (as will be shown) a few years before Uhlenbeck and Goudsmit, with their hypothesis of the electron spin and its anomalous magnetic moment, made it all appear so simple—before quantum mechanics resolved the puzzling replacement of the number J^2 by $J(J + 1)$ that was empirically demanded, and at a time when the vector model of angular momenta was still largely an untried device.

To arrive at his conclusions the moment in time that Landé did called for, not only a special blend of erudition and a brilliant mind, but also the courage to oppose current beliefs and accepted physical principles, even if these were sanctioned by a Niels Bohr, an Arnold Sommerfeld, or a Wolfgang Pauli.

In the development of these prerequisite attributes a benevolent fate lent a helping hand. Landé was fortunate indeed to receive his scientific apprenticeship in the creative and intellectually charged atmosphere that the universities of München and Göttingen provided when the nineteen-year-old Landé enrolled as a student, in 1908, and afterwards. Both universities were centers of learning in mathematics and physics, drawing promising scholars, not only from Germany, but from all parts of the world. In München, where Landé attended his first lectures and was to receive his Ph.D. in 1914, Arnold Sommerfeld presided over a whole school of talented theorists. Led by Peter Debye, Paul Epstein, Peter Ewald, Max von Laue (whose demonstration of the first x-ray diffraction pictures greatly impressed Landé), and others, they were working on the forefront of knowledge, unfolding the early consequences of Planck's amazing quantum postulates.

In Göttingen, where Landé moved on two occasions—the first time, in 1910, to conclude that he should become a theoretical, rather than an experimental, physicist; the second time, two years later, to become *Hauslehrer* to David Hilbert, with the assignment to keep the famed mathematician informed on the latest developments in physics—the mathematical firmament was lit by several stars of the first magnitude. These included, besides Hilbert, Felix Klein, E. Landau, J. Runge, O. Toeplitz, and L. Prandtl (the founder of aerodynamics). Still on the rise were Hermann Weyl, R. Courant, P. Bernays, G. Polya, and other students or instructors destined to achieve fame later on. In the physics department, Woldemar Voigt (famous for his electro- and magneto-optics and Landé's principal physics teacher) and E. Riecke represented the classical school, while the young scholars such as Max Born, E. Madelung, T. von Kármán, E. Freundlich, P. Hertz, and V. Fock were steeped in studies of relativity, quantum theory, and aerodynamics. Moreover, there was, of course, the steady stream of memorable visitors, among them the young Niels Bohr from Copenhagen and, from Leiden, the great Hendrik Lorentz—whose *Theory of Electrons* interested Landé, still in his junior year, so much that he translated it into German, for his own edification. The two centers of the group of young physicists and mathematicians were the *Mathematische Lesezimmer*, with its unique library of pure and applied mathematical literature, and a local café, where meetings took place every afternoon for serious discussions of the latest theories. The prestige that went with being Hilbert's assistant gained Landé access to these circles, as a junior member but with high aspirations of becoming a full-fledged scientist.

The outbreak of World War I suddenly interrupted this promising setup, with Landé managing to obtain his doctorate (under Sommerfeld) just two weeks before the guns started to boom. Joining the Red Cross, Landé served for a few years in eastern-front hospitals, until Max Born succeeded in placing him as his assistant at the Artillerie-Prüfungs-Kommission in Berlin to work on sound-detection methods.

It was while collaborating with Born, towards the end of the war, that Landé received, as he recalls, his first taste of "sharing in overcoming fixed ideas by innovations." In an attempt to derive the bulk properties (such as compressibility) of crystals from the behavior of their constituent atoms, Born had begun to explore the forces responsible for holding the ions together in alkali-halide crystals. His joint labors with Landé led to the unexpected conclusion that Bohr's model of coplanar electronic orbits could not be upheld. Instead Born and Landé had to postulate that the electronic orbits were inclined to one another. Their coauthored paper of 1918, presenting this revolutionary suggestion, caused quite a stir, because, in Landé's words, "the mere analogy of Bohr's orbits with those of the planets around the sun, repeated over and over in the literature, served as a psychological block, for half a dozen years, to a consideration of the possibility of three-dimensional systems."

A method for treating spatial distributions of electrons had already been proposed by Sommerfeld in his famous 1916 paper in *Annalen der Physik*: The quantization rule requiring the angular momentum of a circular orbit to be a multiple of $\hbar$ (Bohr, 1913), had to be generalized to hold for each periodic degree of freedom, in the form $\oint p_i \, dq_i = n_i \hbar$. On application to the elliptical orbits of any electron, assumed to be moving in the Coulomb field of the nucleus alone, these rules were shown by Sommerfeld to imply the spatial quantization of the electron's orbits: The component of the orbital angular momentum in any chosen direction became restricted to values that were integral multiples m (the *equatorial* or *magnetic* quantum number) of the quantum $\hbar$.

However, in planning their attack on the problem of atoms as spatial structures, Born and Landé completely ignored Sommerfeld's space-quantization hypothesis, which was perhaps not so surprising in view of Sommerfeld's own temporary lapse in allegiance to this idea just then. Landé, on Born's suggestion, instead set out to fix the electron configurations in atoms by requiring the electron positions to satisfy the reflection and rotation groups of appropriate polyhedrons. Sommerfeld and other physicists received this symmetric atomic model with great enthusiasm, and Landé published quite a number of papers on this subject after August of 1919. Quantization of the cubic and tetrahedral atomic models and calculation of their energies were followed by a survey to discover the appropriate symmetry for each element in the periodic table (in particular, Landé strongly advocated a tetrahedral arrangement of the four valence electrons of the carbon atom). Although this program failed, it was not without a lasting effect: Through criticism of Landé's carbon atom model, Niels Bohr was led in 1920 to develop his own theory of penetrating orbits, which superseded Landé's theory and subsequently furnished an improved starting point for the study of electronic motions.

Before this took place, however, Landé was obliged temporarily to neglect the problem of symmetrical atoms in order to concentrate intensively on the quantum theory of the helium atom, which was the subject he selected for his *Habilitationsschrift* at the University of Frankfurt-am-Main at the beginning of 1919. The spectrum of helium had been investigated earlier by Friedrich Paschen, but a theory capable of explaining its main features was conspicuously lacking. Especially puzzling was the inferred existence of noncombining singlet and doublet (actually triplet, it turned out later) term systems; it was as if helium consisted of two separate substances (named parhelium and orthohelium).

On the basis of Sommerfeld's hypotheses, the stationary states of an excited helium atom (presumably due to one electron only) had to resemble that of the alkali atoms, with energy levels given by $E = Rhc[n + f(k)]^{-2}$. Herein, n, the *principal* quantum number, stood for the sum $n = k + n'$ of the *azimuthal* and *radial* quantum numbers of the radiating electron; while the function $f(k)$, representing a deviation from the simple Bohr energy formula, took account of the fact that (due to the perturbing action of the inner electron) the outer electron was no longer moving in the simple Coulomb field of the nucleus alone. The various spectral term sequences, labeled S, P, D, etc. (in modern notation) were associated with the values $k = 1, 2, 3$, etc. of the azimuthal quantum number, while the successive terms in a given sequence belonged to the values $n = 1, 2, 3$, etc. for fixed k. The failure of this theory was that it gave no indication of why there should be both singlet and "doublet" term sequences or why, indeed, some terms should have a "doublet" structure. (Today we know, of course, that the parhelium spectrum arises from transitions among singlet, $S = 0$, states, the orthohelium spectrum from transitions among triplet, $S = 1$, states, and that intercombination of terms is forbidden by the selection rule $\Delta S = 0$.)

Predictably (in hindsight), Landé's extensive calculations turned out to be in error (as Wolfgang Pauli, then a twenty-year-old student of Sommerfeld, was quick to point out). But then, it was not the numerical outcome but the impulse given to the conceptual development of quantum theory by some of Landé's novel ideas that came to matter afterwards. Let us briefly recount them.

Having lately come to appreciate the explicit demands of Sommerfeld's quantization rules, Landé proposed to use the discrete orientation possibilities of the plane of the outer electron of helium (assumed to be different from that of its inner electron) to account for the richness of this element's spectral terms. His original approach was to ignore, à la Sommerfeld, the interaction between the angular momenta $\mathbf{L}_o$ and $\mathbf{L}_i$ of the outer and inner electrons, and to quantize the inclination of $\mathbf{L}_o$ relative to the fixed axis in space that $\mathbf{L}_i$ was presumed to furnish. But Landé soon realized that the neglect of the interaction between the electrons was unphysical and that, more realistically, the total angular momentum $\mathbf{L}$ of the atom, not $\mathbf{L}_i$, would define the fixed quantization direction. From this conviction emerged the picture of two electronic orbital planes, together with the vectors $\mathbf{L}_o$ and $\mathbf{L}_i$ perpendicular to them, precessing about the vector $\mathbf{L}$, which remained fixed both in magnitude and direction; a situation which Landé recognized as having a close analog in planetary mechanics. A scrutiny of the principles of quantum theory, as expounded by Bohr (in his book) and J. M. Burgers (in his dissertation), moreover convinced Landé that the magnitude of $\mathbf{L}$, just like that of its components $\mathbf{L}_o$ and $\mathbf{L}_i$, had to be integral multiples of $\hbar$. In the summer of 1919 a synthesis of the foregoing considerations appeared as a brief paper (see the bibliography of Landé's writings), which asserted for the first time the now familiar rules for combining two quantized angular momenta to produce a quantized resultant. The Landé model of the helium atom became the prototype of the general model of angular momenta, which subsequently proved to be a tool of extreme usefulness, both in his own work and that of others.

The helium theory led to Landé's admission as *Privatdozent* at the University of Frankfurt (where Max Born came to occupy the chair of theoretical physics) and earned him an invitation to Bohr's Institute in Copenhagen, where Landé was also to report on his symmetric model of the carbon atom. Bohr encouraged this visit, seeing in Landé's efforts to devise atomic models an overlapping with his own program aimed at determining electronic structures throughout the periodic system of elements. The visit of October 1920 turned out to be beneficial to both parties, with Bohr profiting from criticism of Landé's carbon atom model and with Landé drawing encouragement from Bohr's views on what had become Landé's latest absorbing interest: the *anomalous Zeeman effect.* Bohr, he learned, shared the opinion, also expressed by Sommerfeld (but with less power of persuasion), that the anomalous magnetic-splitting effect concealed some of the profoundest secrets of atomic structure.

At the Physical Institute of the University of Frankfurt, where experiments were in progress on the magnetic splitting of spectral lines as well as of molecular beams (the latter experiments, under O. Stern and W. Gerlach, to culminate in the complete vindication of Sommerfeld's space-quantization hypothesis), the atmosphere was eminently suited to the new direction of Landé's research. It was here that he, in December 1920, commenced his attack on the anomalous Zeeman effect by painstakingly wading through the latest and most comprehensive published survey of this subject, authored by Sommerfeld. Unlike the normal Zeeman effect, which was given its quantum-theoretical explanation four years earlier by Debye and Sommerfeld, the (much more prevalent) anomalous effect had withstood all attempts at theoretical interpretation ever since it was first observed, in 1897, by T. Preston. Because, on the whole, multiplet spectral lines exhibited anomalous splitting, while singlets underwent only the normal effect, it was clear from the start to Landé, especially in view of the Rydberg-Ritz combination principle, that the problem of the existence of the anomalous effect was entwined with that of the multiplet line structure. The solutions of both problems could be expected to emerge together.

Landé discerned a successful approach to this joint problem by combining his earlier adopted vectorial presentation of angular momenta with his so-called *magnetic-core hypothesis* (*Z. Physik*, vols. 5 and 7, 1921). According to this hypothesis, the total angular momentum **J** of an atom had to be viewed as the resultant of the angular momenta **K** and **R** of the optical (outer) electron and the atomic core (German, *Rumpf*), respectively, the core consisting of the nucleus and the remaining (inner) electrons. The interaction between the magnetic moments associated with **K** and **R** caused their precession about **J**, which itself was constant in magnitude and direction. The magnitudes $l\hbar$, $r\hbar$, and $j\hbar$ of **K**, **R**, and **J** were supposed to be quantized, in agreement with the available empirical data, such that $l \equiv k - 1 = 0, 1, 2, \ldots, n - 1$ (n and k, Sommerfeld's principal and azimuthal quantum numbers) and r and j could be integral or half-integral, depending on the multiplicity of the state under consideration. (Actually, Landé operated with the quantum numbers $K = l + \frac{1}{2}$, $R = r + \frac{1}{2}$, $J = j + \frac{1}{2}$, instead of the numbers l, r, j, which we have introduced here to facilitate comparison with quantum-mechanical quantities used earlier.) The different discrete orientations of **K** with respect to **R** would obviously result in different values of j; and this, Landé surmised, was the explanation for the separated sublevels of a spectral term and thus for the appearance of line multiplets consisting of several components.

On application of an external magnetic field **H**, classical electrodynamics predicted that the vector **J** would precess about the field direction and cause the multiplet (term) sublevel in question to suffer a definite displacement ΔE, depending on the orientation of **J** relative to **H**. To specify the possible quantized values, in units of $\hbar$, of the component of **J** along **H**, Landé introduced the quantum number m_j, capable of varying in integral steps over the interval $-j \leq m_j \leq j$. Assigning in this manner a set of m_j values (either all integer or all half-integer) to each unperturbed multiplet sublevel, Landé discovered, by a minute examination of available spectroscopic data, that he could impute the number of lines and their polarizations in Zeeman patterns (of atoms such as sodium) to transitions, obeying the selection rules $\Delta m_j = 0, \pm 1$, between the displaced Zeeman levels. More surprisingly, the energy difference ΔE, separating an arbitrary Zeeman level from its unresolved multiplet sublevel, was correctly given only if the classical expression $m_j \mu_B H$ was multiplied ad hoc by a numerical factor g, whose value showed no variation with m_j but did obviously depend on the quantum numbers l, r, j defining the unperturbed sublevel.

But what, if any, was the unique general formula that encompassed all the separate enigmatic g's, or *splitting factors* (in Landé's terminology)?

The answer to this question became possible in the fall of 1922, when Landé, through the intervention of Friedrich Paschen, was called to Tübingen, the "Mecca of Spectroscopy," as associate professor of theoretical physics. Here, at Paschen's Physical Institute, a large collection of experimental results on Zeeman spectral types, meticulously gathered by Ernst Back over the years, were awaiting theoretical analysis. In the hands of Landé they yielded, after a month of intensive labor, the g formula as well as the interval rule for the separation of multiplet sublevels. Reported in *Z. Physik*, vol. 15 of 1923, these theoretical results agreed with those stated in the first part of our essay, except for the replacement of L, S, J by $K - \frac{1}{2}$, $R - \frac{1}{2}$, $J - \frac{1}{2}$.

In searching for a theoretical explanation of the empirical g formula on the basis of his vector model, Landé pursued essentially the converse argument to our earlier derivation, to come to the following important conclusions. First, the magnitudes of the vectors **K**, **R**, **J** had to be modified through replacement of l^2 by $l(l + 1)$, and similarly for r and j; second, the ratio of the magnetic moment of the atomic core to its angular momentum had to be assigned twice the value predicted by classical reasoning.

The irrationality of the momentum quanta and the anomalous magnetic moment of the core (which was the source of the half-integer m_j values for even-multiplicity states) were vexing riddles which Landé failed to solve. In the meantime, though, the key to the unraveling of the observed Zeeman patterns was found by an astonishing feat of inductive reasoning, and the possibility for predicting even magnetic patterns never before seen was opened up. More important in the long run was the plain fact that Landé succeeded, where Sommerfeld and other great minds failed, only by flouting central precepts of the old quantum theory and by proposing a further invalidation of classical electrodynamics. This realization among physicists no doubt paved the way for the acceptance of the electron spin concept and lent greater urgency to the call for a radically modified quantum-mechanical theory, whose dramatic rise the next few years were to witness.

Landé remained attached to the University of Tübingen until 1931; the years after the *g* formula were taken up by a score of publications on the interpretation of higher spectral multiplets, the derivation of spectral-line intensity rules, and other gaps in the then existing knowledge of atomic theory. Several review articles and books from his pen, dealing with the Zeeman effect, axiomatic thermodynamics, and the new quantum mechanics, also enriched the scientific literature of that era and guided upcoming physicists. During the fall semester of 1929, an invitation to give a series of lectures at Ohio State University—"a most lucky event of my scientific and personal life," as the recipient has described it for various reasons—brought Landé to Columbus, Ohio. A second semester of teaching at O.S.U. in 1930–1931 persuaded him to settle for good in the United States, just two years before the fatal turn in German history.

During the years of adaptation to life in America and of the worrisome march of events in Europe, Landé produced relatively few original papers in the factual sense. The praxis of quantum physics, then reeling under the rapid advances in nuclear and field theory, was beginning to lose a leading innovator. But its pedagogy and the essential analysis of its fundamentals were about to gain, in Landé, an equally undaunted thinker. As time went on, Landé became more and more immersed in the task of organizing, for his own benefit and that of his students, the quantum-mechanical principles put forth in the late 1920s by de Broglie, Born, Heisenberg, Schrödinger, Dirac, and others. The first fruits of these activities were the monographs entitled *Principles of Quantum Mechanics* and *Quantum Mechanics*, published in 1937 and 1950, respectively.

In writing these textbooks, Landé had no difficulty with the rules of quantum-mechanical calculation, about which unanimity existed among scientists, then as now. The stumbling block was to be the attending aspect of a meaningful operational interpretation of this mathematical formalism, about which the agreement was less than universal. Faced with the latter problem, Landé at first elected to follow the *Copenhagen interpretation*, the foundation of which was laid in the famous discussions between Bohr and Heisenberg in Copenhagen during the winter of 1926–1927, and whose essence was contained in Heisenberg's well-known 1927 paper on the indeterminacy principle and in Bohr's highly publicized Como (Italy) address of September 1927 on his own complementarity principle. In espousing the Copenhagen interpretation, Landé was following a course that could count on the consent of the majority of scientists. After all, the Fifth Solvay Congress of 1927 (a few weeks after the Como Congress), where Bohr's ideas were dissected and refined by some of the sharpest minds in physics, seemed to have given the stamp of general approval to the Copenhagen school of thought, and it ended with even Einstein, the foremost dissenter, admitting defeat (although only to remain skeptical the rest of his life). Early misgivings on the Copenhagen interpretation among "less informed" philosophers of science (Karl Popper, as early as 1934, among them) did not seem to merit serious attention among physicists, who as a group were anyway too preoccupied with attacks on practical problems of atomic and nuclear physics to bother much about questions of interpretation.

Gradually, however, the uneasiness which Landé felt in keeping faith with the Copenhagen interpretation in his textbook *Quantum Mechanics*, ripened into complete disenchantment with the logic of this "ideology," its subjective and dualistic modes of interpretation, its stubborn refusal to contemplate possible nonquantal origins of quantum theory, and finally its "idolatory of authorities." The outcome of this development was the publication of the polemic *Foundations of Quantum Theory, a Study in Continuity and Symmetry* (1955), followed by its steadily improved versions *From Dualism to Unity in Quantum Mechanics* (1960) and *New Foundations of Quantum Mechanics* (1965), together with a continuing outpour of papers in

American and European journals, clarifying and amplifying Landé's divergent views on the subject of quantum-mechanical interpretation and related matters.

To appreciate the nature and extent of Landé's break with orthodox opinion, let us briefly recall the main points of the Bohr-Heisenberg arguments underlying the Copenhagen school of thought. The central question around the year 1927 and ever afterwards—prompted by the observed discreteness of physical properties such as energy and angular momentum and by the *dual, wave-particle*, behavior of radiation (cf. photoelectric and Compton effects) and, equally, electrons (cf. diffraction effects)—was this: Assuming the ability of quantum-mechanical theory to encompass all the foregoing phenomena, to what extent, if at all, could one continue to uphold the intuitive classical notions of continuity, causality (or, more precisely, determinism), position, and momentum, and the idea of an external world that existed apart from man's senses?

Heisenberg's partial answer to this compound question is based on his uncertainty principle (first derived with the aid of the Dirac-Jordan transformation theory). Nature, he maintains, presents us only with situations in which the uncertainties (the standard deviations, in mathematical statistics) in the simultaneous values of a coordinate q and its conjugate momentum p are limited by the relation $\Delta p \, \Delta q \geq \hbar/2$. An intuitive understanding of this restriction can be derived from an examination of what happens (or can be imagined to happen) in individual situations: Although, it turns out, either p or q of a particle can be measured with arbitrary precision, any measurement of p involves of necessity an interaction between this system and some measuring device, which gives rise to an uncontrolled and unpredictable change in q; and conversely for a measurement of q.

This restriction on the applicability of the classical notions of position and momentum spells, according to Heisenberg, the downfall of causality as classically understood. For, "In the strong formulation of the causal law, 'If we know exactly the present, we can predict the future,' it is not the conclusion but rather the premise which is false. We cannot know, as a matter of principle, the present in all its details." While the uncertainty principle makes an assertion about prediction, it is emphasized by Heisenberg that it does not concern itself with retrodiction, because, although one might reason that from an accurate measurement of the position of a particle with precisely known velocity precise foregoing (p, q) pairs can be inferred (at least for the instant directly preceding the measurement), these retrodicted values are not susceptible to experimental verification and thus lack any physical reality. A similar insistence on empirical verifiability forces us to abandon the classical habit of speaking of the "true values" of physical quantities such as p and q, since only "blurred" values are within reach even of idealized experiments anyway. The Heisenberg principle, it is suggested, represents truly a property of an atomic particle, not simply a deficiency in our knowledge of the particle.

Niels Bohr was unsympathetic to Heisenberg's formal interpretation of the uncertainty relations (as if nature would "imitate a mathematical scheme") and the remaining bias it bespoke in favor of the intuitive particle concept of classical physics. He pinpointed instead the empirical wave-particle duality as the premise from which any interpretation of quantum theory should commence. The conceptual dilemma, he argued, posed by the Einstein–de Broglie relations, $E = h\nu$ and $p = h/\lambda$, which connect corpuscular attributes with antithetic wave attributes (for radiation exhibiting photonic behavior or electrons simulating de Broglie waves), cannot be resolved simply by appropriate modification and restriction of the classical notion of a particle. Reality is more intangible and requires the use of both particle and wave pictures, in a *complementary* relation to one another, for its full understanding. Admittedly, particles (sharply demarcated in space) and waves (spread out over space) are mutually exclusive concepts, but by always heeding the limitations imposed by the uncertainty relations, we are able to successfully avoid any marriage of these contradictory pictures. The notion of complementarity applies to position q and momentum p: Both these concepts are essential to an exhaustive description of the behavior of an electron, say, but the inverse proportionality linking the accuracies with which they can be known, rules out their incompatible simultaneous use.

Central to Bohr's interpretation is the conclusion that spatio-temporal description of microevents is complementary to their deterministic description: On the one hand, the temporal evolution of the probabilistic wave function ψ is completely determined by Schrödinger's equation; on the other hand, however, the isolation of the system, which this equation presupposes, is broken, and thus causal behavior interrupted, the very instant we attempt to measure the coordinates and momenta required for the space-time description.

Finally, the Copenhagen interpretation declares the ideal of a fully objective description of nature to be unattainable even in principle. We cannot, as is believed possible in classical physics, describe in a completely objective manner what happens in atomic events. For, the wave function, on which hopes for an objective description might fasten, embodies subjective as well as objective features—knowledge of the system peculiar to the observer, together with statements on the definite probabilities of various experimental outcomes. Moreover, as mental analysis of various thought experiments would reveal, we cannot use a given wave function to decide what happens between observations, without falling into glaring logical contradictions. The term "happens" can in fact be applied with confidence only to the observations themselves. The idea of observation is, however, traditionally rooted in a division of the world into observed object and macroscopic observer or observing device. Not only is this division historically arbitrary, but the behavior of the measuring device can be characterized exclusively with the concepts of classical physics, these constituting the language to which evolution has attuned the human mind. Thus, again subjective elements enter into our description of events.

Landé sharply assails the vagueness (praised as flexibility by apologists) of the Copenhagen ideology and heaps ridicule on the flights of metaphysical prose that Heisenberg and other leading Copenhagen theorists, notably L. Rosenfeld, M. Born, C. F. von Weizsäcker, and J. Jeans, occasionally allow(ed) themselves in its attempted elucidation. At the same time, he systematically cuts the ground from underneath every one of the elaborate arguments underlying the Copenhagen "doctrine." While in opposing the Bohr-Heisenberg creed, Landé has eminent company (although not necessarily consistent supporters) among living scientists and philosophers of science (not to mention the antedating skepticism of Planck, Einstein, and Schrödinger), his contribution to the ongoing debate is rare in that it goes well beyond the luxury of rejection and criticism: In the place of every cherished tenet that he irreverently demolishes, Landé submits a counterproposal, compelling in its directness and simplicity.

Landé takes dead aim at the avowed dualism of complementary pictures, the cornerstone of Bohr's approach. There exist no actual alternate corpuscular and wavelike "manifestations" (a dualist term) either of matter or radiation, he emphasizes, although it might sometimes be helpful to reason as if there were. Matter consists of particles, radiation of waves, and everything else ascribed to these two categories of reality is just uncalled for "ideological ballast."

Ironically, this "either-or" *unitary* view of Landé became possible with, and is suggested by, the dualist Max Born's own brilliant probabilistic interpretation (1927) of the Schrödinger wave function. Born's posit is particularly relevant to the proper interpretation of the diffraction patterns exhibited by electrons (or other atomic particles such as neutrons), whose apparent resistance to a corpuscular treatment was historically the main reason why the claim of the wave manifestation of matter was so favorably received by physicists. What the blackening of a photographic plate (or counter readings) in such experiments represents is simply the statistical distribution of individual particle impacts. Each of these particles has a definite position and momentum at any time, as the uncertainty principle must be regarded as referring to the statistical behavior of an assembly of electrons (which is more fully described by the complete wave function), not to any individual member of the assembly. We do not need (here or in any other similar situation) the hypothesis that when the width of an aperture is Δx, every electron passing through it will have a position blurred over a range Δx and a momentum component blurred over the reciprocal range $\Delta p_x \sim \hbar/\Delta x$, as the entrenched interpretation of the uncertainty principle avers. In fact, measurement of the exact p_x value of the electron arriving in the recording area would reveal the accurate value that p_x possessed within the aperture Δx. To Heisenberg's objection that such a value of p_x, like other similar retrodicted property values, is devoid of physical meaning, Landé retorts that if this were so, we would be restrained to speak only of coincidences in space and time. For, are not all other atomic measurements precisely of this indirect type to which Heisenberg objects? (A velocity measurement, for example, requires two position values and two time readings.)

But how is one to explain, on the basis of Landé's unitary particle theory, that the diffraction pattern of electrons impinging on a screen with (say) two openings resembles that of electromagnetic radiation of an appropriate wavelength, or that electrons and x rays are diffracted by crystals into patterns of identical form? In these cases, doesn't the appearance of dark spots in places where any single opening or crystal plane would have guaranteed brightness prove conclusively that we are dealing with typical interference effects peculiar to waves alone? The answer, according to Landé, is a definite no; and if this comes as a surprise, it is the fault of an inexplicable failure of physicists to take note of the pertinent contributions made by W. Duane as early as 1923 and by P. Ehrenfest and P. S. Epstein in the following year.

The key contribution of Duane (*Proc. N.A.S.*, vol. 9, 1923) was to establish parity between linear momentum and the likewise-conserved properties of energy and angular momentum, by supplementing the well-known quantization rules of Planck and of Wilson and Sommerfeld (for energy and the angular momentum conjugate to any periodic coordinate, respectively) by a quantization rule $\int p\, dq = nh$ (n any integer) for the permissible transfer of linear momentum from radiation to a crystal, the integral to be extended over the relevant periodicity interval of the crystal. Landé took this prescription, postulated by Duane in connection with crystal diffraction of x rays (where its use must be regarded as an artificiality), and applied it to the diffraction of electrons instead. For the special case of electrons incident in a plane perpendicular to a set of parallel crystal planes of separation L, the integration interval in question is $(q, q + L)$, and Duane's rule is easily seen to imply that electrons of momentum p will undergo reflections only at the discrete angles θ_n corresponding to the perpendicular momentum change $2p \sin \theta_n = nh/L$. But this result is identical with the Bragg equation derived from the theory of wave interference; and in general Duane's rule makes possible the derivation of von Laue's equations, with their reference to crystal axes rather than atomic planes.

Duane's quantum rule, governing the quantum-mechanical activity of a crystal, thus suffices to explain the selective deflection of electrons entering crystals, without the need to postulate an intermediate wave manifestation obeying the de Broglie relation $\lambda = h/p$. Likewise, the extension of Duane's theory due to Ehrenfest and Epstein (*Proc. N.A.S.*, vol. 10, 1924), ascribes in a natural fashion the diffraction pattern of a screen with two openings (say) to the quantal activity of the diffractor. What happens here is that momentum is imparted by the incident electrons to each one of the three-fold infinity of sinusoidal elementary lattices, of spacings l_1, l_2, l_3 and amplitude $A(\mathbf{1})$, into which the matter distribution of the screen can be analyzed according to Fourier's theorem. The increments $\Delta p_i = h/l_i$, which, like the set of l_i values, form a continuous spectrum, are communicated to the incident electrons with probabilities proportional to $A^2(\mathbf{1})$ for each constituent lattice. The resulting diffraction pattern, generated by the statistical distribution of the individually deflected electrons over the detection area, again resembles in all respects the pattern one would calculate via a wave-theoretical detour. Since the amplitudes $A(\mathbf{1})$ for a screen with one aperture differ from those of a screen with two openings, the diffraction patterns for the two cases are naturally different. Also, the famous question, How can an electron passing through one hole be sensitive to the existence or absence of a second hole, so as to vary its contribution to the diffraction pattern accordingly? acquires a ready explanation: The amplitude $A(\mathbf{1})$, governing the prevalence of the momentum transfer h/l_i, is determined by the matter distribution for the entire screen, not merely for the neighborhood of the point where electron and diffractor interact.

The dogma of duality is equally gratuitous in its application to light or other forms of electromagnetic radiation, where photonic particles are postulated in complementary opposition to waves, with the aid of the Einstein-Compton equations $E = h\nu$ and $p = h/\lambda$. For, according to the modern theory of radiation, advanced in the 1930s by Fermi and Heitler, the reality that confronts us is a continuous electromagnetic field, which can be resolved mathematically into monochromatic waves of definite frequencies and wavelengths: the illusion of photons is created by the further result that the energy and linear momentum of each wave component varies only in discrete multiples of $h\nu$ and h/λ, respectively. It is readily conceded that the photon construct is very useful in the straightforward explication of simple interaction phenomena between radiation and matter, but the same concept must be severely strained in order to accommodate some other effects of electromagnetic radiation, whose understanding in terms of a unitary wave theory offers no difficulty.

In rounding off his case against dualism and the Copenhagen interpretation generally, Landé had to solve the problem of deriving, *ab initio*, the Planck, Wilson-Sommerfeld, and Duane quantum rules, and indeed the whole mathematical edifice of quantum mechanics, without recourse to wave-particle duality or any other facet of the complementarity viewpoint. It seemed obvious to Landé that, with the complementarity principle, also any other quantum principle should be banished from the premises of such a derivation; that the key "expected to unlock the quantum riddle . . . can be found only in a nonquantal basis of postulates which ought to be simple and plausible, so that they do not require further justification."

Landé's efforts to achieve this goal underwent gradual improvement over the years and culminated in a scheme (first presented at an international colloquium in Denver in May 1966, and published in *Physics, Logic, and History*; see bibliography) for the construction of quantum mechanics by way of three nonquantal postulates. In brief, these are: (1) If $P(A, B)$ denotes the probability that an atomic system, originally known to occupy the state A, will afterwards be found (under a B-state test) in a state B, and $P(B, A)$ the probability for the inverse event, then $P(A, B) = P(B, A)$. (2) There exists one general relation among the different $P(A, B)$, which is identical *on the average* with the conventional addition law $P(A, C) = \sum_B P(A, B)P(B, C)$. (3) The connection between time and energy, as well as that between a coordinate q and its conjugate momentum p, is covariant in atomic mechanics, i.e. the connection is independent of the zero points of the quantities involved. Surprisingly, these simple axioms, taken together, give rise to the Born commutation rule, the quantization rules mentioned above, Schrödinger's wave equation, and the uncertainty relations—in short, to the bulk of ordinary quantum mechanics, albeit with an action constant h of undetermined value. Further nonquantal arguments explain moreover why the wave functions of identical-particle systems must be either symmetric or antisymmetric (*New Foundations of Quantum Mechanics*, p. 89).

Two features of Landé's a priori construction of quantum mechanics are especially noteworthy. On the one hand, it is postulated that the states of an atomic system are dynamically interrelated only in a statistical, or probabilistic, fashion. Thus, Landé does not attempt to restore determinism to nature, as some other prominent critics of the Copenhagen philosophy, such as Planck, Einstein, Schrödinger, and de Broglie at one time or another have wished to do. He imposes instead a not-further-analyzed statistical structure on microevents from the very start. On the other hand, while agreeing with the Copenhagen proponents on the demise of determinism, Landé in no way shares their (or, more specifically, Born's) belief that "the appearance of chance in the elementary processes means the end of a sharp separation between the object observed and the subject observing." Neither does he admit the intrusion of subjectivity into quantum physics to be required, any more than in classical physics, by the circumstance that the observer must "decide beforehand which kind of answer he wants to obtain" or by the fact that "the means of observation depends on the subject" (quotations of Born).

Landé's reinterpretation of the basic concepts of quantum mechanics and his reconstruction of its mathematical formalism have come under attack by Born and Heisenberg (*Physik. Blätter*, vol. 25, 1969) and have been discussed and reviewed by several other scholars, including one of us (W. Yourgrau, *Brit. J. Phil. Sci.*, vol. 12, 1961). The introduction to a *Festschrift* is not the proper place to reiterate the arguments that have or could be raised pro and contra Landé or to resume the debate, which will no doubt continue. Suffice it to stress here a point on which critics and supporters of Landé's views could concur: Landé, by the comprehensible eloquence of his assault on entrenched shibboleths and the attractiveness of his counterproposals, has probably done more than any living scientist to make the rank and file of physicists and philosophers wary of imbued beliefs and—through his admonition that the taken-for-granted Copenhagen philosophy does not represent a logically necessary conclusion—aware of the mutability of the foundations of physics. In having this effect, Landé has rendered a service to the future of science and its foundations that transcends the aims of his particular *cause célèbre*.

Ever since his student days at the University of Münich, where he witnessed the unavailing efforts of young physicists to explain energy quantization by searching for a defect in the ergodic theorem of classical mechanics, Landé has been obsessed with the "quantum riddle," with the idea of reducing it to less strange origins. Dissatisfied with the radical departure from intuition that the generally accepted interpretation of quantum mechanics demanded, the unshackled and maverick mind of Landé—which years earlier recognized a simple scheme where only confused magnetic line patterns met the "fixed stare" of his contemporaries—made a discovery of simplifying impact: In wave-particle duality, scientists were badly confusing a mere analogy of behavior with something worthy of profound thought, a conception of what a particle is with what its statistical behavior may suggest. This discovery marked the beginning of a long road of perseverance that finally led to the demonstration "that quantum theory, instead of being a set of enigmatic, though most successful, rules of calculation, can be understood as the logical consequence of a few almost self-evident postulates of symmetry and invariance." The quantum riddle has found its solution, at least to Landé's own scientific conscience.

What more than the fulfillment of a self-imposed program, denied to great men like Maxwell and Einstein, could the true scientist ask for, unless it is the wide discussion and appreciation of his fondest ideas in his own lifetime? That this, too, will befall Alfred Landé, is the essence of our good wishes to him on this congratulatory occasion.

Wolfgang Yourgrau — Denver, Colorado
Alwyn van der Merwe — January 1971

Perspectives in Quantum Theory

One **Max Born**

An Open Letter to Alfred Landé

My dear Landé,

My age prohibits me from contributing a scientific paper to this volume in your honor. But I do not wish to be excluded from those who congratulate you by an article, and I beg you to allow me to send you a short letter with a few reminiscences of our contacts in life.

My memory is now bad, and I can hardly recollect anything from the time when you came to Göttingen as a student and attended classes of mine, except that I soon discovered your abilities and high standards and came to know you a little.

Later, during the First World War, you were, like me, attached to the Prussian military administration called A.P.K. (Artillerie-Prüfungs-Kommission), in particular to the section led by Rudolf Ladenburg. We were occupied with *Schallmessverfahren* (called sound ranging by the British), which was an acoustical method to determine the position of guns by measuring the time of arrival of their reports at different (at least three) stations of observation. We both soon found this work rather dull and began to do scientific research, using one drawer of our desk for the sound-ranging paper, the other for our physical calculations—we were convinced that our chief, Herr von Jagwitz, would not be able to distinguish between the two kinds of hieroglyphics.

There we spent many days working out the formula for the energy of ionic crystal lattices, much assisted by our colleague Madelung, who found a way to overcome the difficulty which arose from the "bad" convergence of the series representing the electrostatic part of the energy. I wonder whether one incident will be remembered by you as vividly as it is by myself. All went well with the calculations; we got excellent agreement between the theoretical and the observed lattice constants and compressibilities for simple ionic lattices, when we assumed that each ion was a system of coplanar rings as Bohr or Sommerfeld had suggested. I gave the manuscript to Einstein to present to the Berlin Academy for publication. But one morning, entering our room in the A.P.K., I found you in desperation: you had discovered that we had made an elementary mistake in the calculation of the repulsive forces as double sums over all pairs of ions. We had forgotten the factor $\frac{1}{2}$ necessary to avoid counting each pair twice. I checked your statement and had to admit that our paper was wrong. There was nothing to do but to ask Einstein (who lived quite near to the A.P.K. building) to withdraw our paper. I had to perform this sad duty and informed Einstein about our mistake. He was not shocked, but instead laughed and said that our reasoning seemed to him essentially sound, and that we should revise our paper and would certainly obtain satisfactory results in the end. So we began to work again, generalizing our assumptions by dropping the hypothesis that the electrons in ions were arranged in Bohr's ring-shaped orbits, and assumed that they had a higher spatial symmetry.

This led to success, and we could publish our results. Looking through these old papers again, I have noticed that this failure of the ring models was even at that early time recognized by me as an indication that Bohr's much admired theory was leading in wrong directions; and this together with similar results encouraged me to look for a new quantum mechanics, which was, ten years later, discovered in collaboration with Heisenberg and Jordan.

In the intermediate time the foundations of the quantum hypothesis was secured by many deep considerations and surprising discoveries. While I did not take part in this work, you were one of the leaders in this era of physics. I need not enumerate your many fundamental results which were obtained by an incredible amount of industry and patience combined with ingenious guessing and strict calculations.

For some period we were separated, until the years 1920–1921. I was professor in Frankfurt when you appeared there to spend some time in my tiny restful department, to finish some of your amazing work on multiplet spectral lines and their Zeeman effect. If I remember correctly, you were sitting for some weeks at my desk, opposite to me, and were deeply immersed in your calculations, while in another room Stern and Gerlach performed their celebrated experiments.

I recall only one other time that we met. It was during the well-known Volga Congress (in 1928, I think). This started in Leningrad, but was moved to Moscow, where the main and longest discussions were held. Joffe, who had initiated this conference and was its leader, performed miracles in the art of translation, repeating every lecture, given in a foreign language, in Russian for his Soviet colleagues, and vice versa. From Moscow we traveled to Nizhni Novgorod. There the whole congress was transferred to a Volga steamer, and here the main section of the conference took place. We traveled down the Volga her whole length, stopping at many towns like Kazan, Saratov, and others, where we were taken by cars or buses to the biggest hall of the place and there had to lecture and discuss physical problems. You had come all the way from Tübingen, and we spent a jolly time with one another.

I remember hardly anything of the scientific side of the "traveling congress" but solely the impressive landscape around the enormous river and the picturesque towns and villages on its banks. I was overtired all the time and sometimes a little hungry; the young Russian students of both sexes were very gay and sang, danced, and flirted all day and night. There was poor service on board; since no free Soviet citizen would accept (at that time, just after the revolution) a permanent job as waiter, cook, or steward, there existed no trained people in these professions, and some men had to be picked out arbitrarily to do the necessary chores on the boat. So it happened that each meal was some hours late and lasted a very long time. I recollect that once you and I were offered a whole sturgeon for dinner; we accepted, paid a small extra sum, invited a few colleagues, but then had to wait for hours until we were served—but we enjoyed this excellent fish.

A night I remember particularly well was when you replaced the young noisy Russians at the piano and played Chopin on the old bad instrument so beautifully that I have never forgotten it. I hope you have continued playing and still find pleasure and consolation in your music.

I have to say a word more though: a word of thanks. You were wise enough to leave Germany when the stench of the coming nationalism and anti-Semitism was still faint and to settle in America. I had no such opportunity and suffered the fate of many other refugees. We were driven out and went to South Tyrol; the only comfort in our distress was the marvelous mountain scenery—until your letter arrived in which you offered me a position at your university in Columbus, Ohio. I cannot describe to you the joy and relief we felt when we saw that we were not forgotten. Later several other similar invitations came, and we chose in the end Cambridge in England because I had been there as a student and knew the English way of life. Also, at that time the Cavendish Laboratory under Rutherford in Cambridge was a center of physics. I hope you have not blamed me for this decision. In any case, your invitation was the first one and therefore the most welcome one.

Now I end my letter of reminiscences with my most cordial congratulations and good wishes,

Max Born

Two **Louis de Broglie**

A New Interpretation Concerning the Coexistence of Waves and Particles

1. Introduction

At the moment when the friends and admirers of Alfred Landé are celebrating his eightieth birthday, I am happy to join in the homage rendered to this great physicist who is the author of numerous and important works and whose name remains attached to one of the most celebrated formulas of contemporary theoretical physics. I am, moreover, all the more happy owing to the fact that Landé has during recent years directed very lively and pertinent critiques at the interpretation of formalisms currently employed in the usual investigations of quantum physics—an interpretation which, resulting from the ideas of the Copenhagen school, has become today for the majority of theoretical physicists, despite the objections of Einstein and Schrödinger, the sole "orthodox" doctrine.

That orthodox interpretation was not at all what I had in mind in 1923–1924 when I arrived at the idea forming the basis of wave mechanics: that the notion of the coexistence of particles and waves extends to all particles.[1] This coexistence had been discovered by Einstein in 1905 for light in his theory of photons or light quanta. Pursuing the same course, I had been led to envisage, under the name of "the theory of double solution," an interpretation of wave mechanics as set forth in an article in the *J. Phys. Radium* of 1927.[2] Afterward, the difficulty of working out my position completely and diverse other circumstances persuaded me for a long while to give my adherence, though often a cautious one, to the Copenhagen school's position. However, about fifteen years ago, I resumed the further development of my original ideas and have written a great number of articles, a list of which one may find in the references. I have been assisted in this task by only a few young researchers, whom I feel obliged to thank for being engaged along with me in this arduous endeavor.

These publications comprise a critical part setting forth evidence of the difficulties encountered in the actual interpretation of quantum physics. This critical part is quite close to Landé's ideas on the subject. But my articles also contain a constructive part conforming to my original notions and quite different from Landé's envisaged solutions. I am not able to reproduce here the critical part of my thoughts, but refer any curious readers to the works cited in the references. I shall content myself with resuming the general line of the constructive part of my work by recalling the arguments that enabled me originally to discover wave mechanics—ideas never mentioned in an exact fashion, as the ordinary textbooks and other expositions regrettably show.

2. The Origins of Wave Mechanics

When, in 1922–1923, I had my first ideas about wave mechanics, I was guided by the vision of constructing a true physical synthesis, resting upon precise concepts, of the coexistence of waves and particles. I never questioned then the nature of the physical reality of waves and particles.

One thought struck me forcefully. The phase of a plane monochromatic wave, given by the expression

$$2\pi \left(\nu t - \frac{\alpha x + \beta y + \gamma z}{\lambda} \right),$$

allows one to define a four-dimensional vector of a space-time "wave" with components $\nu/c, \alpha\nu/V, \beta\nu/V, \gamma\nu/V$, where V is the phase speed. This shows that the wave frequency ν is transformed like energy by way of the formula $\nu = \nu_0/(1 - \beta^2)^{1/2}$ whereas the frequency of a clock transforms according to the formula $\nu = \nu_0(1 - \beta^2)^{1/2}$ as follows from the relativity theory of clock retardation. I noticed afterwards that the above four-vector, defined as the space-time gradient of the phase of a plane monochromatic wave, could be written as the energy-momentum four-vector of a particle if one introduced the Planck constant h and set

$$W = h\nu \qquad \text{and} \qquad p = \frac{h}{\lambda}, \qquad [2.1]$$

in order to connect the energy W with the frequency ν, on the one hand, and the magnitude of the linear momentum p with the wavelength λ, on the other. I was also led to suppose that the particle, located constantly at one point within the wave, possesses both this energy and momentum; further, that the particle describes one of the rectilinear rays of the wave.

This is, however, never recalled in the standard expositions of quantum mechanics. I also noticed that if the particle was regarded as containing the rest energy $M_0c^2 = h\nu_0$, it was natural to compare it with a small clock of proper frequency ν_0, such that, when the clock moves with the speed βc, its internal frequency differs from that of the wave and equals $\nu_0(1 - \beta^2)^{1/2}$. I then easily demonstrated that during the motion which I attributed to the particle in the wave, the particle has an internal vibration which remains constantly in phase with that of the wave.

The account given in my *Thèse* had the disadvantage of applying only to the particular case of a plane monochromatic wave, which is never rigorously realized in nature, by virtue of the inevitable existence of spectral width. I saw clearly that if a complex wave can be represented by a Fourier integral, i.e. by a superposition of components, then these components exist only in the mind of the theorist; and, insofar as these components have not been separated by a physical process destroying the initial superposition, it is the superposition itself which is the physical reality. It then occurred to me, shortly after my *Thèse*, to generalize the ideas which guided me in this work to distinguish the real physical wave of my theory from the fictitious wave ψ (in the statistical sense and arbitrarily normalized) which physicists, following the results of Erwin Schrödinger and Max Born, began to introduce systematically into the investigations of wave mechanics, but which was not the wave I had initially conceived of. My reflections also encouraged me to investigate, in an article in *J. Phys. Radium* of May 1927 under the title "Theory of the Double Solution," a new interpretation of wave mechanics for the case of a general wave, and moreover to generalize the law of particle motion, which I had until then envisaged only for the too restricted case of the plane monochromatic wave.

3. The Theory of the Double Solution

I do not intend to examine here in detail the theory of the double solution and shall therefore limit myself to referring the reader to the investigations which I have published on the subject for about fifteen years.[3] I shall indicate only the two general ideas upon which the theory rests:

a. The wave is, according to my notions, a physical wave of very weak amplitude whose essential role is to guide the motion of strong local concentrations of energy constituting the particles. The wave may not be arbitrarily normalized and is therefore distinct from the wave ψ, of statistical nature, utilized in quantum mechanics. I designate by v the physical wave and, in order to recover the statistical sense of the wave ψ, I define the latter by the relation $\psi = Cv$, where C is a normalization factor. It is this essential distinction between the two solutions v and ψ of the wave equation which prompted me to name my theory the "theory of the double solution."

b. For me, the particle, localized always within its wave as time proceeds, forms in the wave a very small region of high energy concentration, which one may, in a first approximation, regard as a shifting singularity. Considerations, which would be too long to reproduce here, have led me to recognize that the motion of a particle must be defined in the following manner. If the complex solution of the wave equation, which describes the wave v (or, if one prefers, the wave ψ, which amounts to the same thing in virtue of the relation $\psi = Cv$), is written in the form

$$v = a(x, y, z, t)e^{(i/\hbar)\varphi(x,y,z,t)}, \qquad [2.2]$$

where a and φ are real functions, then the energy W and the momentum $\mathbf{p}$ of the particle, when it occupies the position x, y, z at the instant t, are given by the formulas

$$W = \frac{\partial \varphi}{\partial t} \qquad \text{and} \qquad \mathbf{p} = -\nabla \varphi. \tag{2.3}$$

This furnishes, in the case of the plane monochromatic wave, the expressions [2.1], but Eq. [2.3] can also be applied to any wave whatsoever. If in the formula [2.3] one writes W and $\mathbf{p}$ in the form

$$W = \frac{M_0 c^2}{\sqrt{1 - \beta^2}} \qquad \text{and} \qquad \mathbf{p} = \frac{M_0 \mathbf{v}}{\sqrt{1 - \beta^2}},$$

$\mathbf{v}$ being the particle's velocity with $\mathbf{v} = \beta c$, one obtains

$$\mathbf{v} = \frac{c^2 \mathbf{p}}{W} = -c^2 \nabla \varphi \left(\frac{\partial \varphi}{\partial t} \right)^{-1}. \tag{2.4}$$

We may call Eq. [2.4] the "formula for the particle's guidance by the wave." It is easily generalized when the particle is submitted to an external field.

Let us now introduce the ideas at the origin of wave mechanics, according to which, we recall, the particle may be compared to a small clock of proper frequency $\nu_0 = M_0 c^2 / h$; and let us give it a velocity $\mathbf{v}$ defined by the formula [2.4]. For an observer who sees the particle passing with a speed βc, the internal frequency of this clock is $\nu = \nu_0 (1 - \beta^2)^{1/2}$. This allows one to demonstrate readily that, even in the general case of a wave which is not plane monochromatic, the internal vibration of the particle remains constantly in phase with the wave carrying it. This result, which is the same as in the particular case of the plane monochromatic wave, can be considered as the essential content of the guidance law expressed by [2.3] and [2.4], and it constitutes our most general statement.[4]

The theory shows then that the rest mass M_0 which appears in the expressions for W and for $\mathbf{p}$ is not in general equal to the rest mass m_0 usually attributed to the particle. Indeed, one finds that, instead,

$$M_0 = m_0 + \frac{\delta q_0}{c^2}, \qquad [2.5]$$

where δq_0 is a positive or negative variation of the rest energy of the particle. The quantity q_0 is the "quantum potential" of the double solution theory; its value depends on the amplitude of the wave function. In the writings listed in Ref. 1, there is an expression for q_0 pertaining to an electron, which may be used to represent its wave, either with the Klein-Gordon relativistic equation or with Schrödinger's equation (the Newtonian devolution of the former), or finally with Dirac's more complete equations. The expression for a photon is also given, should one employ the Maxwell equations, complemented with a very small mass term, when one calculates the electromagnetic energy.

4. Comparison of a Particle's Rest Energy to Hidden Heat

About 1908, Planck and von Laue demonstrated that if a body absorbs a quantity of heat Q_0 as measured in a stationary system of reference, then the quantity of heat that the body conducts to its interior relative to a coordinate system with respect to which it has a constant velocity $\mathbf{v} = \beta c$ is given by

$$Q = Q_0 \sqrt{1 - \beta^2} \qquad [2.6]$$

Because of the invariance of the entropy, this leads to the temperature transformation law $T = T_0(1 - \beta^2)^{1/2}$.

Formula [2.6], accepted without question for a long period of time, has recently been questioned by numerous authors. As for myself, after long reflection on this problem, which is both difficult and susceptible to confusion, I have arrived at the conclusion that formula [2.6] is really exact and, in particular, that it is without question applicable to a very small body such as a particle. For a detailed exposition of my views regarding this issue, see the appropriate papers in the references.[5]

In recent years, since 1961, the similarity between the relativistic transformation formulas of a clock frequency, $\nu = \nu_0(1 - \beta^2)^{1/2}$, and of the heat amount, $Q = Q_0(1 - \beta^2)^{1/2}$, has appeared very important to me for the following reason: If one compares a particle to a small clock of proper frequency $\nu_0 = M_0c^2/h$, it is a logical step to conceive of the particle's rest energy M_0c^2 as a heat amount Q_0 concealed within the particle's interior. Indeed, a clock contains within itself energy of internal agitation, not associated with any total momentum—which corresponds to the most general definition of a body containing heat in a state of equilibrium.[6]

If then $Q_0 = M_0c^2$ is the caloric energy of a particle in its proper system, the heat that is conducted toward its interior, in the reference system where the particle moves with a velocity of $\mathbf{v} = \beta c$, is given by

$$Q = Q_0 \sqrt{1 - \beta^2} = M_0c^2 \sqrt{1 - \beta^2} = h\nu_0 \sqrt{1 - \beta^2}.$$

The particle thus appears to us to be simultaneously like a small clock of frequency $\nu = \nu_0(1 - \beta^2)^{1/2}$ and a small heat reservoir of content $Q = Q_0(1 - \beta^2)^{1/2}$ in motion with a velocity $\mathbf{v} = \beta c$. And it is solely the identity between the transformations for the heat amount and for the frequency of a clock which makes this double aspect possible.

5. The Hidden Thermodynamics of Particles

Relativistic dynamics teaches us that the Lagrangian function of a particle not subjected to any external field, whose rest mass is M_0 and whose velocity $\mathbf{v} = \beta c$, is $\mathscr{L} = -M_0c^2(1 - \beta^2)^{1/2}$. By definition, the action integral is

$$\int \mathscr{L}\, dt = \int -M_0c^2 \sqrt{1 - \beta^2}\, dt = -\int M_0c^2\, dt_0,$$

t_0 being the particle's proper time. The action being obviously invariant, it is tempting to establish a relation between the two fundamental invariants—action and entropy; but to make this possible, one must assign a well-defined value to the action integral by choosing its integration interval of time conveniently. With the ideas already in our possession, it is natural to identify this interval with the period T_i of the particle's internal vibration, for the usual rest mass m_0, with respect to the frame of reference in which the particle speed is βc. Since $T_i^{-1} = (m_0 c^2/h)(1 - \beta^2)^{1/2}$, one may also define a cyclical action integral, after noticing that (the period T_i being always very small) M_0 and β can be treated as constants within the interval of integration, by

$$\mathscr{A} = -\int_0^{T_i} M_0 c^2 \sqrt{1 - \beta^2}\, dt = -\frac{hM_0}{m_0}. \qquad [2.7]$$

We are now able to define the entropy of the particle by

$$\frac{S}{k} = \frac{\mathscr{A}}{h}, \qquad [2.8]$$

where k and h are respectively Boltzmann's and Planck's constants. Since $Q_0 = M_0 c^2 - m_0 c^2$, [2.7] and [2.8] yield

$$\delta S = -k \frac{\delta Q_0}{m_0 c^2}. \qquad [2.9]$$

We have therefore arrived at a point where we may attribute to the particle's motion within its wave a certain entropy and, as a result, a certain probability P given by Boltzmann's formula $S = k \ln P$.

From these ideas I have been able to derive two results which appear to me to be very important[6]: (1) The principle of least action is only a particular case of the second law of thermodynamics. (2) The privileged status (whose paradoxical character Schrödinger has stressed) that present-day quantum mechanics has bestowed upon plane monochromatic waves and stationary states of quantized systems may be explained by the fact that they correspond to maximum entropy; other states are not nonexistent but merely less probable.

6. The Hidden Thermostat

The thermodynamic concept of a particle that we have investigated suggests that a particle, even when isolated from a complete macroscopic body, is constantly in thermal contact with a kind of thermostat residing in what we shall call the void.

We have compared a particle to a body containing heat, since we regard it as a system possessing internal vibrational energy without total momentum. But, this system being very simple, it appears difficult to attribute entropy and temperature to it. On the contrary, the hidden thermostat is, by definition, a very complex system, and one is prone to recognize that it furnishes heat to a particle when the rest mass of the latter increases with a positive variation of quantum potential, $\delta Q_0 = c^2 \, \delta M_0$, while the thermostat recovers heat when the particle's mass decreases with a negative variation of quantum potential. As in Einstein's theory of Brownian movement, the entropy defined by the formulas [2.8] and [2.9] must be that of the hidden thermostat; and the quantity δq_0 is, in the particle's proper frame, both the variation of quantum potential and the variation δQ_0 of the particle's internal heat. One is then justified to write, in the frame of reference where the particle's speed is βc,

$$\delta S = - \frac{\delta Q_0 \sqrt{1 - \beta^2}}{T_0(1 - \beta)} = - \frac{\delta Q}{T} . \qquad [2.10]$$

Herein T_0 is defined by

$$T_0 = \frac{h\nu_0}{k} = \frac{m_0 c^2}{k} , \qquad [2.11]$$

and the minus sign is explained by the fact that heat is lost by the hidden thermostat whose entropy is S.

One should not be too surprised to see that the temperature, which the hidden thermostat seems to have, depends upon the particle's rest mass m_0. Indeed, it is through the intermediary effect of the wave v, whose structure depends upon m_0, that the particle is in contact with the hidden thermostat.

The great progress realized in thermodynamics at the time when the molecular structure of matter and statistical mechanics were introduced has made possible the recognition that when a body is, to all appearances, in a stable thermodynamic state, it continually undergoes in reality small fluctuations, averaging zero, about that state. It is in this way that one has been able to develop a theory of fluctuations and of Brownian movement. We should expect to encounter analogous circumstances in our hidden thermodynamics.

In order to study these questions, I refer once again to my publications cited in Ref. 6 and restrict myself to recalling the following point. I have demonstrated in my book treating the hidden thermodynamics of particles[7]—in order to justify the well-established fact, at least for the Schrödinger equation, that the expression $|\psi^2|\, d\tau$ yields the "probability of presence" of a particle within the element of volume $d\tau$ at the instant t—the necessity of admitting that the particle jumps continually from one of its guidance trajectories to another. The trajectory defined by the guidance law is then only a "mean" trajectory; it would be described in reality if the particle did not undergo continual perturbations, which result from the aleatory heat exchanges with the hidden thermostat.

A simple comparison will render the preceding more understandable. Let us consider the hydrodynamic flow of a fluid. A granule placed in the fluid finds itself carried along by the fluid motion. If the granule is heavy enough not to feel appreciably the action of the individual shocks that it receives from the invisible molecules of the fluid, it will describe one of the streamlines of the hydrodynamic flow. If one compares the granule with a particle, the assembly of fluid molecules is comparable to the hidden thermostat of our theory, and the streamline described by the granule is the guidance trajectory. However, if the granule is sufficiently light, its motion will continually be disturbed through the individual chance impacts with the molecules of the fluid. It will then be animated by a Brownian movement, causing it to jump constantly from one streamline to another. The same is true for a particle in thermal contact with a hidden thermostat; and thus one may obtain a theory of Brownian movement for a particle in its wave (see my paper of April 10, 1967, cited in Ref. 6).

7. Conclusion

After having indicated the general lines of the reinterpretation of wave mechanics that we are pursuing, it is interesting to note the directions in which wave mechanics can be developed. These appear to be the following four: (1) reinterpretation of the usual formalism of quantum mechanics; (2) development of the hidden thermodynamics of particles; (3) reinterpretation of the quantum theory of fields; (4) theory of elementary particles.

My associates and I are actively pursuing our investigations in these diverse directions. Without giving the bibliography of all that we have already published, I shall simply say that, in view of the small number of researchers who up to now have shown interest in my ideas, the progress appears encouraging.

My age scarcely allows me to hope to see the culmination of these efforts, but I am convinced today that the precise physical ideas investigated in this article can lead in the future, in agreement with a new theory of the particle structure, not only to a reconsideration of the formalisms of present-day quantum mechanics and quantum field theory but also to the opening up of entirely new perspectives in quantum theories generally.

References

Selected Writings of Louis de Broglie

1.
Compt. Rend. **177**, 506, 548, 630 (Sept.–Oct. 1923). *Thèse de Doctorat* (Masson et Cie., 1924). Reedited 1963.

2.
J. Phys. Radium, Series 6, **5**, 225 (1927).
3.
Une tentative d'interprétation causale et non linéare de la Mécanique ondulatoire: la théorie de la double solution (Gauthier-Villars, 1956); English translation: *Non-linear Wave Mechanics: a Causal Interpretation* (Elsevier, 1960). *La théorie de la Mesure en Mécanique ondulatoire* (Gauthier-Villars, 1957). J. Phys. Radium **20**, 963 (Dec. 1959). *Etude critique des bases de l'interpretation actuelle de la Mécanique ondulatoire* (Gauthier-Villars, 1963); English translation: *The Current Interpretation of Wave Mechanics: a Critical Study* (Elsevier, 1964). *Certitudes et incertitudes de la Science* (Albin Michel, 1966). *Ondes électromagnetiques et photons* (Gauthier-Villars, 1968).
4.
J. Phys. Radium **28**, 481 (May–June 1967).
5.
Compt. Rend. **263**, 1351 (Dec. 1966); **264**, 1173 (Apr. 1967).
6.
La Thermodynamique de la particule isolée (ou Thermodynamique cachée des particules) (Gauthier-Villars, 1966). Ann. Inst. Henri Poincaré **1**, No. 1, 119 (1966). Compt. Rend. **257**, 1822 (Sept. 1963); **264**, 1041 (Apr. 1967).
7.
La Thermodynamique de la particule isolée (ou Thermodynamique cachée des particules) (Gauthier-Villars, 1966), pp. 80–87.

Three **Hans-Jürgen Treder**

The Einstein-Bohr Box Experiment

In the discussions on a possible fusion of quantum field theory and general-relativistic gravitational theory, *Gedankenexperimente* play a special role for the reason that, at present, real possibilities for experimenting in this area do not exist and mathematical constructs with clear physical significance are not available. An often discussed group of *Gedankenexperimente* of this kind leads, for example, to the conviction that the consideration of general-relativistic effects in quantum theory (or of quantum effects in general relativity theory) yields the smallest measure for length (and, along with that, time) measurements. This limiting measure is given by the quantity $l = (hf/c^3)^{1/2}$, "Planck's elementary length," which Planck introduced in 1906 on the basis of dimensional analysis; it is just that length which can be constructed from the three fundamental constants the speed of light c, the action quantum h, and the gravitational constant f.

While this group of *Gedankenexperimente* are to some extent perspicuous in the physical sense,[1] another *Gedankenexperiment*, when taken seriously in the interpretation current up till now, has a disquieting consequence, as Landé points out in his book *New Foundations of Quantum Mechanics.*[2] Landé treats the so-called "Einstein-Bohr box experiment" and points out that the familiar interpretation given by Bohr[3] of Einstein's *Gedankenexperiment* (involving the precise determination of energy changes through the process of weighing) would have as a general result that quantum mechanics must be logically dependent upon relativistic gravitational theory. For, on principle, Heisenberg's fourth uncertainty relation is, according to Bohr, compatible with the functioning of a spring balance only by virtue of the Einstein clock-retardation effect in a gravitational field. Landé remarks that this paradoxical conclusion speaks against Bohr's interpretation of the Einstein box experiment.

In opposition to this, Rosenfeld emphasized, in his extensive discussion of the relation between quantum theory and gravitation,[4] that Bohr discovered only an inconsequential result of Einstein's chain of thought, since Einstein's proposal to use a spring balance to measure energy presupposes both of his equivalence principles —the equivalence between energy and inertial mass and the equivalence between inertial and gravitational mass—and that the second of these includes even the Einstein time dilation. Rosenfeld explains further that the gravitational field in the Einstein box experiment plays only an intermediary role and that what matters basically is the determination of the impulse exchange between the spring and the contents of the box. We will see further below that the role of the gravitational field in the box experiment is indeed completely incidental, from which it also follows that the appeal to Einstein's time dilation for the explanation à la Bohr of this experiment does not correspond to the real state of affairs.

Landé has in fact observed that, according to the Einstein equivalence principles, the time dilation brings about the proportionality between displacements of energy, time, and place (ΔE, ΔT, and Δq):

$$-\frac{\Delta E}{E} = \frac{\Delta T}{T} = gc^{-2}\,\Delta q, \qquad [3.1]$$

so that from a latitude ΔT of the time measurement resulting from Δq, a corresponding quantal reciprocity of the latitudes of energy and time cannot be derived.

In order to enter into a discussion, let us recall the arguments of Einstein and Bohr included in Ref. 3. Questioning the general validity of Heisenberg's fourth uncertainty relation $\Delta E\,\Delta T \gtrsim h$, Einstein maintained that it is possible, on the basis of the equivalence principles, to determine the energy content E of a box through weighing by means of a spring balance; this requires a measurement of the corresponding extension q of the spring (which, for purposes of simplicity, is assumed to obey Hooke's law with a spring constant α). The box has a hole in its side, which Einstein supposes can be opened or closed by a shutter that is governed by a clock. At two definite and predetermined times T_1 and T_2 the shutter is automatically opened and then closed. If during the time $T = T_2 - T_1$ any particle escapes from the box and thereby changes its energy content, then, according to Einstein, the precision of the determination of the time of this change is evidently independent of the precision with which one weighs the energy content. Hence, this instant in time may be determined with an accuracy that is not limited by the Heisenberg uncertainty relation.

To this reasoning Bohr raised the objection that the reading of the length q of the spring is possible only with an uncertainty Δq given by

$$\Delta q \gtrsim \frac{h}{\Delta p}, \qquad [3.2]$$

and the minimum uncertainty Δp in the momentum of the box "must obviously . . . be smaller than the total impulse which, during the whole interval T of the balancing procedure, can be given by the gravitational field to a body of mass Δm." Hence,

$$\frac{h}{\Delta q} \sim \Delta p < Tg\,\Delta m = Tgc^{-2}\,\Delta E, \qquad [3.3]$$

wherein g denotes the acceleration due to gravity. Bohr concludes: "The greater the accuracy of the reading q of the pointer, the longer must, consequently, be the balancing interval T, if a given accuracy Δm of the weighing of the box with its content shall be obtained."

Now, from general relativity theory, a displacement of the box by an amount Δq in the direction of the gravitational field causes a change ΔT in the clock reading in accordance with Einstein's formula for the red shift in a gravitational field:

$$\Delta T = Tgc^{-2}\,\Delta q. \qquad [3.4]$$

Therefore, from the uncertainty Δq of the position reading, on the one hand, and the condition [3.3], on the other, precisely the Heisenberg uncertainty relation

$$\Delta E \, \Delta T \gtrsim h \qquad [3.5]$$

is obtained.

The critical element in Bohr's argument is, in my opinion, the relationship between the uncertainty in the momentum of the box and the duration T of weighing. We shall return to this point later. At present we want to show, through a somewhat modified Einstein experiment, that Bohr's assumption, namely that the consideration of the time dilation [3.4] provides the explanation of the Einstein box experiment, cannot be correct. We assume that the box is, for practical purposes, massless and contains indistinguishable electrically charged particles, say, protons, which may be at rest with respect to one another and to the observer. Furthermore, the protons may be sufficiently far apart, so that the total energy E of the box is given by the sum of the equilibrium energies of the individual protons: $Mc^2 = \sum m_p c^2 = E$.* Since all protons possess the same specific charge e/m_p, an equivalence principle holds; in the situation described here, it expresses the proportionalities between the total charge Q, the total energy E, and the total mass M:

$$Q = \frac{e}{m_p} M = \frac{e}{m_p} \frac{E}{c^2}. \qquad [3.6]$$

We suspend the box, prepared in this manner, in a homogeneous electrostatic field which is produced, for example, by a negatively charged plane surface. The field strength of this field is then given by $\mathbf{F} = (F, 0, 0)$, and we find a proportionality of the spring's elongation q to the total charge Q of the box:

$$\alpha q = QF. \qquad [3.7]$$

* One can also replace the protons by suitable macroscopic particles, e.g. by equal-sized charged particles or bubbles, where—as in the Millikan experiment—each drop contains exactly one elementary charge. In this case, the determination of the mass through "electrical weighing" is feasible not only in the sense of a *Gedankenexperiment*.

On account of [3.6], this produces also a proportionality between the spring elongation and the energy content of the box.

If corresponding application of Bohr's considerations are made here, we obtain the relationship between the uncertainties Δp, Δq, and ΔE (of the box-momentum, elongation, and energy), on the one hand, and the duration T of the weighing process, on the other:

$$\frac{h}{\Delta q} \sim \Delta p < TF\,\Delta Q = \frac{e}{m_p c^2}\,TF\,\Delta E. \qquad [3.8]$$

Here there is, however, no longer any Einstein time dilation, that is, the uncertainty Δq of the elongation causes no uncertainty ΔT of the time interval T.

Hence, we are able to imagine that, during the time the box is opened, some protons escape from the box (with very small velocities), where for weighing in a homogeneous electrostatic field no relationship exists between the uncertainties of the time and energy measurements. Thus, if Bohr's vindication of Heisenberg's fourth uncertainty relation proves to be correct for a gravitational field, then the uncertainty relation must be invalid for measurements in an electrostatic field.

In fact, however, Bohr's estimate of the uncertainty of the spring's elongation is not understandable, for the uncertainty of the box-momentum is brought about by the intervention of the observer in his attempt to read the spring length q, and this disturbance by the observer has nothing to do with the effect of the gravitational field (or electrostatic field) upon the box. Hence, no connection can exist between the duration of the weighing and the disturbance of the momentum. It is also, in fact, not understandable why the momentum should become more imprecise with increasing weighing time. The Einstein question consequently must be thought through anew, since Bohr's attempt to push it *ad absurdum* has not met with success.

Einstein's assumption, namely, that it is possible to determine at precise points in time the initial as well as the final energy, is of course correct. The question is only how close to one another the measurement times can be moved. To be sure, as has become evident through the discussions of Bohr and Heisenberg, the relation $\Delta E \, \Delta T \gtrsim h$ does not at all assert a connection between the uncertainties of energy and time measurements upon stationary systems; the relation indicates rather that a time interval $\Delta T \gtrsim h/\Delta E$ must separate the first and second measurements in a determination of an energy difference $E_1 - E_2$ with an uncertainty ΔE. Consequently, the fourth uncertainty relation has direct significance only for nonstationary systems, in that it connects the lifetime (or half-life) ΔT of the nonstationary state with its spread in energy (line width), such that if the initial condition of the nonstationary system is precisely given, then the energy of the final state exhibits a line width

$$\Delta E \sim \frac{h}{\Delta T}. \qquad [3.9]$$

This means the following:

Let us consider an ensemble $\sum$ of certain quasi-stationary systems which initially possess a sharply defined energy E_1. Then, in the final state, the energies of the systems constituting the ensemble will be statistically distributed with a half-width given by [3.9]. The final energy E_2 of each individual system is precisely measurable; but the systems possess different final energies, despite their having the same initial energy. (They have then also given off different energy quanta.)

Such a quasi-stationary state now also leads to the uncertainty relation for the Bohr-Einstein box. Let us consider an ensemble of these boxes, all possessing at the start the same spring tension and total energy. Before the emission of a particle from the box occurs, the system "box + spring" is in a state of stationary equilibrium. At the instant that a particle of mass *dm* is emitted, the system "box + spring" does not at first achieve a stable condition, because the spring tension and the gravitational force no longer balance one another. The system begins to vibrate, and indeed with an amplitude dq proportional to the change of the force acting upon the spring

$$\alpha\, dq = g\, dm; \tag{3.10}$$

and it becomes in this manner a quasi-harmonic oscillator with the vibrational energy

$$\tfrac{1}{2}\alpha\,(dq)^2 = g\, dm\, dq. \tag{3.11}$$

As long as the box vibrates up and down, the change dq of the spring elongation and, therefore, $dm = c^{-2}\, dE$ are not readable. The system is, however, not a true harmonic oscillator; rather, the vibrations are damped, while in the spring, through the action of internal friction, and the vibrational energy [3.11] is converted into heat energy, which finally leaves the system through heat conduction or radiation. This heat transfer is a dissipative process in which information is unavoidably lost: The energy carried off is not precisely determined, rather, it spreads statistically, so that each system of the ensemble suffers a slightly different energy loss. Leaving details aside, it follows from the general characteristics of damped vibrations that the fluctuation ΔE of the total amount of dissipated energy of the ensemble members are inversely proportional to the median lifetime ΔT of the oscillation state: $\Delta E \sim h/\Delta T$.

After complete cessation of the vibrations, it is possible to read off accurately the spring elongation and hence the energy content of the system. The individual systems of the ensemble have, however, in the final state somewhat different energies, as each of them has lost a different amount of energy in the damping of its vibrations. Consequently, the different systems of this ensemble will not settle down in the steady state with equal elongations; these will spread. Corresponding to the different energy contents of the boxes, we have

$$\alpha\, \Delta q = gc^{-2}\, \Delta E = \frac{gh}{c^2\, \Delta T}. \tag{3.12}$$

The spread Δq is larger, the shorter the average lifetime of the quasi-stationary vibrational state, and the connection between lifetime and energy spread is given by the Heisenberg uncertainty relation.

Einstein's paradox may thus be resolved in the following manner: With the emission of a particle from the box, the Einstein system "box + spring" goes over into a quasi-vibrational state and through damping of these oscillations evolves into the final, steady state. The energy of the final state scatters from one experiment to another, and the spread ΔE is larger, the shorter the time that the system requires on the average for attaining the final condition. Therefore, in the process of measuring the energy content, a time interval of at least $\Delta T = h/\Delta E$ must separate the initial and final states if the determination of the final state is to result with a reproducible accuracy ΔE. The fulfillment of the fourth uncertainty relation is therefore a consequence of the quantum properties of the spring, and has nothing to do with any connection between quantum theory and gravitation theory.

In conclusion, I wish to thank A. Landé and L. Rosenfeld for our discussions concerning Einstein's paradox and Bohr's proposed solution, as well as O. Singer for his criticism of the manuscript.

References

1.
Cf., for example, H.-J. Treder, "Die Quantentheorie des Gravitationsfeldes und die Plancksche Elementarlänge," in *Physikertagung, Plenarvorträge,* Stuttgart, 1965; Monatsber. Deut. Akad. Wiss. Berlin **8**, 311 (1966).
2.
A Landé, *New Foundations of Quantum Mechanics* (Cambridge University Press, 1965), pp. 122 and 123.
3.
N. Bohr, "Discussion with Einstein on Epistemological Problems in Atomic Physics," in *Albert Einstein: Philosopher-Scientist,* ed. by P. A. Schlipp (Tudor, 1951).
4.
L. Rosenfeld, "Quantentheorie und Gravitation," in *Entstehung, Entwicklung und Perspektiven der Einsteinschen Gravitationstheorie,* ed. by H.-J. Treder (Akademie Verlag, 1966).

Four

Eugene P. Wigner

Quantum-Mechanical Distribution Functions Revisited

I.

There seems to be a revival of interest in the formulation of quantum mechanics in terms of distribution functions, i.e. functions $P(q, p)$ of positional and momentum coordinates q and p. The function P represents a quantity which is, in the classical limit, the phase-space distribution function; it gives the probability that the coordinates and momenta have the values q and p. The existence and properties of such distribution functions are closely related to the question of the possibility of the reformulation of quantum mechanics in terms of classical concepts and are, therefore, close to Landé's interests. The exact definition of P when the quantum conditions, in particular the uncertainty relations, play a role is not clear. One wishes to define, for every normalized state vector ψ, a corresponding distribution function P. This is postulated to satisfy certain conditions which the classical phase-space distribution also satisfies, but these conditions do not determine P completely. The choice between the P which satisfy a set of specified conditions is made aiming at greatest mathematical simplicity and in such a way that the P chosen becomes a useful tool for carrying out calculations of quantities which are not easily obtainable otherwise.

Two reasonable conditions which can be imposed on P are: (a) that it be a Hermitian form of $\psi(x)$, i.e. that the P which describes the same state as the wave function ψ be given by

$$P(q, p) = (\psi, M(q, p)\psi), \qquad [4.1]$$

where M is a self-adjoint operator depending on p and q and (b) that P, if integrated over p, give the proper probabilities for the different values of q as

$$\int P(q, p)\, dp = |\psi(q)|^2, \qquad [4.2a]$$

and, if integrated over q, give the proper probabilities for the momentum as

$$\int P(q, p)\, dq = (2\pi\hbar)^{-1} \left| \int \psi(x) e^{-ipx/\hbar}\, dx \right|^2 . \qquad [4.2b]$$

All integrations, unless otherwise noted, are to be extended over the whole range of the variables, from $-\infty$ to $+\infty$. It may be observed also that, even though the conditions which P should satisfy are formulated above for a one-dimensional problem, they can be easily extended to an arbitrary, many-dimensional one by making q and p vectors with as many components as has the configuration space of the ψ for which P should give an alternate description.

A somewhat milder form of condition (b) is that P should give the proper expectation values for all operators which are sums of a function of p and a function of q as

$$\iint P(p, q)[f(p) + g(q)]\, dp\, dq = \left(\psi, \left[f\left(\frac{\hbar}{i}\frac{\partial}{\partial x}\right) + g(x)\right] \psi\right). \qquad [4.2]$$

It will suffice to use condition (b) in the form [4.2].

A third, very natural, condition on P would be that P is nonnegative for all values of p and q:

$$P(p, q) \geq 0. \qquad [4.3]$$

It has been stated in an article by the present writer where a P satisfying conditions (a) and (b) was given[1] (actually for the many-dimensional case) that these conditions are incompatible with [4.3]. It has been variously suggested to the writer that this proof be published, and this will be the first subject of the present article. It is sufficient, for this purpose, to consider the one-dimensional case.

In order to carry out the proof it will be shown that the assumption that a P satisfying all three conditions [4.1], [4.2], and [4.3] can be defined for every ψ leads to a contradiction. Actually, in order to obtain the contradiction, it will suffice to consider such ψ which are linear combinations $a\psi_1 + b\psi_2$ of any two fixed functions such that ψ_1 vanishes for all x for which ψ_2 is finite. We start with the following lemma:

LEMMA 1

If $\psi(x)$ vanishes in an interval I, and if $g(q)$ is zero outside this interval, and nowhere negative therein, one has for the P corresponding to the ψ above

$$\int P(q, p)g(q)\, dq = 0 \qquad [4.4]$$

for all p (except for a set of measure zero). This follows from [4.2] with $f = 0$: the integral of [4.4] with respect to p,

$$\iint P(q, p)g(q)\, dp\, dq = [\psi, g(x)\psi] = 0, \qquad [4.4a]$$

vanishes because the right side of [4.2] vanishes. However, the integrand with respect to p, that is the left side of [4.4], is non-negative for the g postulated as long as [4.3] holds for P. It follows that the integrand with respect to p must vanish except for a set of p of measure zero. Since, furthermore, [4.4] is valid for every function $g(q)$ which satisfies the conditions of Lemma 1, we can conclude in a similar way that

LEMMA 2

If $\psi(x)$ vanishes in an interval I, the corresponding $P(q, p)$ vanishes (except for a set of measure zero) for all values of q in that interval.

Let us consider now two functions $\psi_1(x)$ and $\psi_2(x)$ which vanish outside of two nonoverlapping intervals I_1 and I_2, respectively. Because of [4.1], the distribution function $P_{ab}(q, p)$ which corresponds to $a\psi_1 + b\psi_2$ will have the form

$$P_{ab}(q, p) = |a|^2 P_1 + a^* b P_{12} + ab^* P_{21} + |b|^2 P_2. \qquad [4.5]$$

Setting $b = 0$, we note that P_1 is the distribution function for ψ_1 and similarly P_2 is the distribution function for ψ_2. Let us consider

[4.5] for the q outside the interval I_1. Since P_1 vanishes almost everywhere for such q, the distribution function [4.5] cannot be positive for all a and b unless both P_{12} and P_{21} vanish (except for a set of measure zero in q and p) if q is outside I_1. The same can be concluded as long as q is outside I_2. Hence, we have, instead of [4.5], almost everywhere

$$P_{ab}(q, p) = |a|^2 P_1(q, p) + |b|^2 P_2(q, p). \qquad [4.6]$$

This means that the distribution function P_{ab} is (almost everywhere) independent of the complex phase of a/b. This seems, at least intuitively, absurd if P_{ab} is to give the proper momentum distribution, i.e. is to satisfy [4.2b].

In order to demonstrate this, let us denote the Fourier transforms of ψ_1 and ψ_2 by $\varphi_1(p)$ and $\varphi_2(p)$. Equation [4.2b] then reads

$$|a|^2 \int P_1(q, p)\, dq + |b|^2 \int P_2(q, p)\, dq$$
$$= |a|^2 |\varphi_1(p)|^2 + |b|^2 |\varphi_2(p)|^2 + 2Re\, ab^* \varphi_1(p) \varphi_2(p)^*. \qquad [4.7]$$

Since this must be valid for all a, b, we must have identically in p

$$\varphi_1(p)\varphi_2(p)^* = 0. \qquad [4.7a]$$

This is, however, impossible since φ_1 and φ_2, being Fourier transforms of functions restricted to finite intervals, are analytic functions (in fact, entire functions) of their arguments and cannot vanish over *any* finite interval. This then completes the proof that no nonnegative distribution function can fulfill both postulates (a) and (b).

II.

Since a bilinear distribution function which gives the proper probabilities for position and momentum (i.e. satisfies the two conditions of the preceding section) cannot be everywhere nonnegative, this last requirement will be abandoned for the rest of this article.[2] The question then arises of what other natural requirements on the distribution function can be made and to what degree these determine the correspondence between wave functions and distributions. When investigating this question, it will be convenient to relax the mathematical rigor of the preceding section. In particular, the operator in (I) will be replaced by a kernel so that we will have

$$P(q, p) = \iint K(q, p; x, x')\psi(x)^*\psi(x')\, dx\, dx'. \tag{4.8}$$

The kernel K will be permitted to be a singular function, though.

The first and most natural requirement on the correspondence [4.8] is that it be Galilei invariant. This means, in one dimension, first, that it be displacement invariant

$$P(q + a, p) = \iint K(q, p; x, x')\psi(x + a)^*\psi(x' + a)\, dx\, dx'.$$

Substituting x and x' for $x + a$ and $x' + a$ on the right side, and considering that [4.8] must be valid for all a, one easily infers that K can depend, in addition to p, only on the differences $q - x$ and $q - x'$, or on $2q - x - x'$ and $x - x'$, not on all three quantities q, x, x'. This was to be expected.

The second requirement of Galilei invariance is that if $\psi(x)$ is replaced by $\psi(x)e^{ip'x/\hbar}$, the new distribution function becomes $P'(q, p) = P(q, p - p')$

$$\begin{aligned} P'(q, p) &= \iint K(q, p; x, x')\psi(x)^*\psi(x')e^{ip(x'-x)/\hbar}\, dx\, dx' \\ &= P(q, p - p') = \iint K(q, p - p'; x, x')\psi(x)^*\psi(x')\, dx\, dx'. \end{aligned} \tag{4.9}$$

Since this must be valid for all ψ, we have

$$K(q, p; x, x')e^{ip'(x'-x)/\hbar} = K(q, p - p'; x, x'). \tag{4.10}$$

Substituting $p' = p$, one finds

$$\begin{aligned} K(q, p; x, x') &= e^{ip(x-x')/\hbar}K(q, 0; x, x') \\ &= e^{ip(x-x')/\hbar}K^0(2q - x - x', x - x'). \end{aligned} \tag{4.11}$$

Since K depends only on the differences $2q - x - x'$ and $x - x'$, a function K^0 of these can be substituted for $K(q, 0; x, x')$. We now have

$$P(q, p) = \iint K^0(2q - x - x', x - x')e^{ip(x-x')/\hbar}\psi(x)^*\psi(x')\,dx\,dx'. \qquad [4.12]$$

We shall find, next, the restrictions on K^0 implied by [4.2] and [4.2a]. We have, according to [4.2a],

$$\int P(q, p)\,dp$$
$$= \iiint K^0(2q - x - x', x - x')e^{ip(x-x')/\hbar}\psi(x)^*\psi(x')\,dx\,dx'\,dp$$
$$= 2\pi\hbar \int K^0(2q - 2x, 0)|\psi(x)|^2\,dx = |\psi(q)|^2, \qquad [4.13]$$

from which it follows that

$$2\pi\hbar K^0(2q - 2x, 0) = \delta(q - x), \qquad [4.14]$$

or

$$K^0(q, 0) = \frac{1}{\pi\hbar}\,\delta(q). \qquad [4.14a]$$

Similarly, [4.2b] gives

$$\int P(q, p)\,dq$$
$$= \iiint K^0(2q - x - x', x - x')e^{ip(x-x')/\hbar}\psi(x)^*\psi(x')\,dx\,dx'\,dq$$
$$= (2\pi\hbar)^{-1} \iint \psi(x)^*e^{ipx/\hbar}\psi(x')e^{-ipx'/\hbar}\,dx\,dx', \qquad [4.15]$$

from which it follows that

$$\int K^0(2q - x - x', x - x')\,dq = (2\pi\hbar)^{-1}, \qquad [4.16]$$

or

$$\int K^0(q, x)\,dq = (\pi\hbar)^{-1}. \qquad [4.16a]$$

We can next introduce the requirement that the correspondence [4.8] be invariant with respect to space and time reflections. The former requirement demands that a simultaneous change of sign of all the variables of K leave it invariant. This means that K^0 is left unchanged if the signs of both of its variables are changed. Time-inversion invariance demands that replacement of ψ by ψ^* change p into $-p$. It follows that

$$K(q, -p; x, x') = K(q, p; x', x) = K(q, p; x, x')^*. \qquad [4.16b]$$

The last part of [4.16b] follows from the Hermitian nature of K, required to give a real P. The first part of [4.16b] renders K^0 even in its second variable—it is therefore even in both of its variables. The second part of [4.16b] renders K^0 real. Therefore, K^0 must be a real and even function of both of its variables. Evidently, these conditions do not fully determine K^0 so that a great deal of arbitrariness remains in the definition of P.

A further condition which can be imposed is that the correspondence between P and the wave function be symmetric in p and q (apart from the sign of i), so that one has, in addition to [4.12], also

$$P(q, p) = \iint K(p, q; k, k')^* \varphi(k)^* \varphi(k')\, dk\, dk', \qquad [4.17]$$

where

$$\varphi(k) = (2\pi\hbar)^{-1/2} \int \psi(x) e^{-ikx/\hbar}\, dx. \qquad [4.17a]$$

This condition will be satisfied if

$$\iint K(q, p; x, x') \psi(x)^* \psi(x')\, dx\, dx'$$
$$= \iiiint K(p, q; k, k')^* (2\pi\hbar)^{-1} \psi(x)^* \psi(x') e^{i(kx - k'x')/\hbar}\, dx\, dx'\, dk\, dk'. \qquad [4.18]$$

It follows that

$$2\pi\hbar K(q, p; x, x') = \iint K(p, q; k, k')^* e^{i(kx-k'x')/\hbar}\, dk\, dk', \qquad [4.18a]$$

or, with [4.11], and since K^0 is real

$$2\pi\hbar K^0(2q - x - x', x - x')e^{ip(x-x')/\hbar}$$

$$= \iint K^0(2p - k - k', k - k')e^{i[q(k'-k)+kx-k'x']/\hbar}\, dk\, dk'. \qquad [4.19]$$

Introducing $\rho = 2p - k - k'$ and $\sigma = k - k'$ as integration variables and writing ξ for $2q - x - x'$ and η for $x - x'$, the equation simplifies greatly, and one has

$$K^0(\xi, \eta) = (4\pi\hbar)^{-1} \iint K^0(\rho, \sigma)e^{-i(\rho\eta+\sigma\xi)/2\hbar}\, d\rho\, d\sigma. \qquad [4.20]$$

This means that K^0 belongs to the characteristic value 1 of the operator Q with the kernel

$$Q(\xi, \eta; \rho, \sigma) = (4\pi\hbar)^{-1}e^{-i(\xi\sigma+\eta\rho)/2\hbar}. \qquad [4.21]$$

The operator Q shows great similarity to that of the Fourier transformation. In particular, its square transforms a function $f(\rho, \sigma)$ into $f(-\rho, -\sigma)$, so that, if f is even in ρ, σ,

$$K^0 = (1 + Q)f \qquad [4.22]$$

will satisfy [4.20]. One can verify, furthermore, that if $f(\rho, 0) = \delta(\rho)/\pi\hbar$, then Qf satisfies [4.16a], and conversely, if f satisfies [4.16a] so that $\int f(\rho, \sigma)\, d\rho = (\pi\hbar)^{-1}$, then $Qf(\xi, 0) = \delta(\xi)/\pi\hbar$. Hence, if f satisfies both conditions, the same will be true for Qf.

It follows that the condition of the symmetry between q and p, as expressed by [4.18], still leaves a great deal of freedom in the choice of K. This is essentially restricted to one-quarter of a Hilbert space.

The original choice of the correspondence between P and the wave function was[3]

$$P(q, p) = (\hbar\pi)^{-1} \int \psi(q + x)^* \psi(q - x) e^{2pix/\hbar}\, dx. \qquad [4.23]$$

This corresponds to

$$K^0(q, x) = (\hbar\pi)^{-1}\, \delta(q), \qquad [4.23a]$$

and this is, evidently, the simplest way to satisfy all the preceding requirements. However, this is not the only possible choice: another, relatively simple selection is

$$K^0(\xi, \eta) = \frac{1}{\pi^2\hbar} \frac{|\eta|}{\xi^2 + \eta^2}, \qquad [4.24]$$

leading to

$$\begin{aligned} P(q, p) &= \frac{1}{\pi^2\hbar} \iint \frac{|x - x'| e^{ip(x-x')/\hbar}}{(2q - x - x')^2 + (x - x')^2} \psi(x)^* \psi(x')\, dx\, dx' \\ &= \frac{1}{2\pi^2\hbar} \iint \frac{|x - x'| e^{ip(x-x')/\hbar}}{x^2 + x'^2} \psi(q + x)^* \psi(q + x')\, dx\, dx'. \end{aligned} \qquad [4.25]$$

III.

The question now arises whether there is any simple criterion which permits a unique coordination of distribution functions to wave functions. The only one known to this writer is to demand that, in the force-free case, the equation of motion of P be the classical one

$$\frac{\partial P}{\partial t} = -\frac{p}{m} \frac{\partial P}{\partial q}. \qquad [4.26]$$

This, together with the postulate of invariance [4.11], and the conditions (a) and (b) of the first section specify P uniquely. The corresponding K will turn out to be given by [4.23a] so that the q-p symmetry condition [4.18] is automatically satisfied.

One finds from

$$\frac{\partial\psi(x)}{\partial t} = \frac{\hbar i}{2m}\frac{\partial^2\psi}{\partial x^2} = \frac{\hbar i}{2m}\psi'' \qquad [4.27]$$

and [4.12] that

$$\frac{\partial P(q,p)}{\partial t} = \frac{\hbar i}{2m}\iint K^0(2q - x - x', x - x')e^{ip(x-x')/\hbar}$$

$$[\psi(x)^*\psi''(x') - \psi''(x)^*\psi(x')]\, dx\, dx'. \qquad [4.28]$$

Partial integrations with respect to x and x' in the first and second terms, respectively, give

$$\frac{\partial P(q,p)}{\partial t} = \frac{\hbar i}{m}\iint\left(2K_{12}^0 + \frac{2ip}{\hbar}K_1^0\right)e^{ip(x-x')/\hbar}\psi(x)^*\psi(x')\, dx\, dx', \qquad [4.29]$$

where the lower indices of K^0 denote differentiations with respect to the first and second variables, respectively,

$$K_{12}^0(\xi,\eta) = \frac{\partial^2 K^0(\xi,\eta)}{\partial\xi\,\partial\eta} \qquad K_1^0(\xi,\eta) = \frac{\partial K^0(\xi,\eta)}{\partial\xi}. \qquad [4.29a]$$

The variables of K^0 in [4.29], however, remain $2q - x - x'$ and $x - x'$. On the other hand,

$$-\frac{p}{m}\frac{\partial P(q,p)}{\partial p} = -\iint\frac{p}{m}2K_1^0 e^{ip(x-x')/\hbar}\psi(x)^*\psi(x')\, dx\, dx' \qquad [4.30]$$

gives just the second term of [4.29]. It follows that

$$K_{12}^0 = \frac{\partial^2 K^0(\xi,\eta)}{\partial\xi\,\partial\eta} = 0, \qquad [4.31]$$

or $K^0(\xi,\eta) = f(\xi) + g(\eta)$. Condition [4.16a] now shows that $g(\eta)$ must be independent of η, and it can be assumed to be zero since a constant can be absorbed into f. Then, [4.14a] shows that $f(\xi) = \delta(\xi)/\pi\hbar$ so that we recovered [4.23a], the old distribution function.

Naturally, in the presence of forces, the distribution function does not satisfy the classical equation of motion

$$\frac{\partial P}{\partial t} = -\frac{p}{m}\frac{\partial P}{\partial q} + \frac{\partial V}{\partial q}\frac{\partial P}{\partial p}, \qquad [4.32]$$

V being the potential. There are many typical quantum phenomena which are in conflict with [4.32]. However, the departure from [4.32] contains at least the second power of $\hbar$. The full equation of motion was given before.[4]

In addition to the correction terms to [4.32], there is a departure from classical theory inasmuch as not every $P(q, p)$ is realizable. Thus, one has, in addition to the normalization integral

$$\iint P(q, p)\, dq\, dp = 1, \qquad [4.33]$$

the added condition

$$\iint [P(q, p)]^2\, dq\, dp = 2\pi\hbar, \qquad [4.33a]$$

which is an expression for the uncertainty relation. In the case where P represents a mixture rather than a pure state, the left side of [4.33a] is actually smaller than $2\pi\hbar$. Equation [4.33a] is a special case of a more general one[5] which gives the transition probability between two states, ψ_1 and ψ_2, in terms of the corresponding distribution functions

$$\left|\int \psi_1(x)^*\psi_2(x)\, dx\right|^2 = (2\pi\hbar)^{-1} \iint P_1(q, p)P_2(q, p)\, dq\, dp, \qquad [4.34]$$

reemphasizing the fact that the distribution functions have to be able to assume negative values. Equation [4.34] is a consequence of the unitary nature of the kernel $(2\pi\hbar)^{-1/2}K$ given in [4.23a] for the old distribution function.

References

1.
Some of the papers dealing with this subject are: E. Wigner, Phys. Rev. **40**, 749–759 (1932); J. E. Moyal, Proc. Cambridge Phil. Soc. **45**, 99–124 (1949); H. Mori, I. Oppenheim, and J. Ross, *Studies in Statistical Mechanics*, ed. by J. De Boer and G. E. Uhlenbeck (North-Holland, 1962), Vol. I, Part C, pp. 213–298; W. E. Brittin and W. R. Chappell, Rev. Mod. Phys. **34**, 620 (1962). This article also contains more complete references to the earlier literature; J. C. T. Pool, J. Math. Phys. **1**, 66–76 (1966); L. Cohen, J. Math. Phys. **1**, 781–786 (1966); K. Imre, E. Özizmir, M. Rosenbaum, P. F. Zweifel, J. Math. Phys. **8**, 1097–1108 (1967). Optical phenomena were treated by means of similar concepts by R. J. Glauber in *Fundamental Problems in Statistical Mechanics*, ed. by E. C. D. Cohen (North-Holland, 1968), Vol. II, pp. 140–187; and J. R. Klauder and E. C. G. Sudarshan, *Fundamentals of Quantum Optics* (W. A. Benjamin, 1968). This book also contains extensive references to earlier literature; G. S. Agarwal and E. Wolf, Phys. Rev. D **2**, 2161–2225 (1970); G. Nienhuis, J. Math. Phys. **11**, 239 (1970).
2.
Positive (more precisely: nonnegative) distribution functions were defined, however, by K. Husimi, Proc. Phys. Math. Soc. Japan **22**, 264 (1940); Y. Kano, J. Math. Phys. **6**, 1913 (1965). See also a forthcoming paper by K. E. Cahill and R. J. Glauber in the Phys. Rev., and W. A. Smith and E. P. Wigner, Bull. Am. Phys. Soc. **14**, 59 (1969). See also F. Bopp, Ann. Inst. Henri Poincaré **15**, 81 (1956).
3.
Ref. 1.
4.
Ibid.
5.
This formula must have been known rather generally, but no explicit reference to it has so far been found.

Five

James L. Park and
Henry Margenau

The Logic of Noncommutability of Quantum-Mechanical Operators—and Its Empirical Consequences

We wish to honor Alfred Landé by scrutinizing in this article one of the shibboleths of the quantum doctrine: the impossibility of performing simultaneous measurements of noncommuting observables. In his book[1] Landé regards as a half-truth the proposition: p and q cannot be measured simultaneously. The present paper presents an examination and indeed a justification of this claim.

1. The Compatibility Problem

Quantum physics, in using *operators* instead of functions to represent observables, complicates the relationship between its observables (i.e. their mathematical representatives) and the empirical numbers to which they must ultimately refer. Perhaps the most controversial difficulty associated with this operator-observable correspondence arises from the commutative law of arithmetic, viz. if a and b are numbers, then $ab = ba$. Naturally this law applies to all measurement results, quite independently of the theory by which they are interpreted. In quantum theory, however, the associated pairs of Hermitian operators do not necessarily commute.

Understandably, the presence in quantum theory of noncommuting observables has from the beginning elicited a great deal of curiosity. Some kind of physical interpretation must be given; the fact, for instance, that $[X, P] \neq 0$ surely expresses something very interesting about position and momentum. But what? The orthodox answer is this: Noncommuting observables are *incompatible*, that is, it is impossible to perform upon a single system *simultaneous measurements* of two such observables. The present paper is an abbreviated report of a systematic analysis of this famous principle of impotence; but first, as a prelude to the substance of the article, we feel it appropriate to review briefly the more common—and frequently illogical—arguments typically advanced in behalf of the doctrine in question. Many of them have already been subjected to criticism by Landé.

The typical historical account of quantum theory from Planck to the present endeavors to present a rather smooth transition from the "old quantum theory" (Bohr atom, particulate photon, classical ontology) to the "new quantum theory" (state vectors, probability, complementarity). Yet any discussion about modern quantum theory which employs concepts peculiar to the "old" to demonstrate alleged features of the "new" is of little value. The following sections therefore contain a *logical* study of the notion of compatibility entirely within the axiomatic framework of (new!) quantum theory, independently of whatever dreams, intuitions, or *Gedankenexperimente* historically might have inspired its ingenious creators.

(1) UNCERTAINTY PRINCIPLE. Many *Gedankenexperimente* have been designed to illustrate Heisenberg's famous law; unfortunately, the false impression is often conveyed that this principle, which is actually a theorem about standard deviations in collectives of measurement results, imposes restrictions on *measurability*. Simple common-sense arguments quite unrelated to the quantum theory could easily be adduced to show the elementary absurdity of such an inference.

(2) PROJECTION POSTULATE (NAIVE VERSION).* Frequently appended to the useful postulates of quantum mechanics is one which, if it were correct, could lead to the incompatibility doctrine as a theorem. It is the notion of wave-packet reduction, according to which any measurement invariably leaves a system in such a state that an immediate repetition of the measurement would yield the same result as the first measurement. The reasoning is: If simul-

* The fundamental irrationality, together with the mathematical strangeness, of the view that a single observation shall in general fix the probability distribution (state vector) of an entire ensemble has been emphasized repeatedly by one of the present authors. (Refs. 4 and 5.) This point is further elaborated in Refs. 2 and 3.

taneous measurement of noncommuting observables were possible, such an act could leave a system in a nonexistent state. This argument is, however, unworthy of serious consideration, since the idea of wave-packet reduction does not survive close scrutiny.[2–5]

(3) PROJECTION POSTULATE (VON NEUMANN'S MEASUREMENT INTERVENTION TRANSFORMATION).[6] There is a way[7] to express the projection postulate in terms of ensembles and the selection of subensembles which does make sense. If this version represented a *universal* trait of measurement, then it would imply the incompatibility principle as a theorem. We have proved this elsewhere.[8] However, it can be demonstrated that even this "reasonable" variant of the projection postulate does not describe *all* physical measurements and is therefore unacceptable as a universal quantal axiom.

(4) PROBLEMS CONCERNING JOINT PROBABILITIES. If joint (i.e. simultaneous) measurements are possible, then there must exist joint probability distributions. However, attempts to generate such distributions for noncommuting observables using fairly standard mathematical ideas have been unsuccessful, and this failure has been interpreted as proof of the incompatibility principle. This position has been examined carefully in another paper by the present writers.[9] There this first probability argument in behalf of the incompatibility principle is traced to the same fundamental errors which underlie the following argument (5). Since arguments (4) and (5) stand or fall together, we shall here concentrate our attention on (5).

(5) VON NEUMANN'S SIMULTANEOUS MEASURABILITY THEOREM. In his classic work on quantum mechanics, von Neumann proved a theorem which provides the best defense ever given of the incompatibility doctrine. Strangely enough, it is also the most widely ignored argument for incompatibility, even though, unlike (1)–(4), it is a logical deduction from a seemingly reasonable quantum axiom set which does not include the projection postulate.* (Cf.

* To be sure, the projection postulate does appear in von Neumann's book, but it plays no role in the theorem here considered.

Sec. 4.) The ensuing sections will emphasize argument (5), the only extant evidence for incompatibility which is firmly embedded in the basic mathematical structure of modern quantum theory.

Because (5) arises deep in the theoretical framework of quantum mechanics, it seems desirable here to interpolate a brief survey of basic quantum axiomatics in order to furnish a basis for distinguishing clearly which quantum statements are hypotheses and which ones are *derivable* propositions. Only in this way can the deduction in (5) be properly evaluated.

As everywhere else, the objects of study in quantum mechanics are called *physical systems.* Associated with them are the constructs known as *observables,* which in turn are correlated via rules of correspondence to empirical operations that generate numbers. These operations are *measurements.* The numbers they produce are called *measurement results,* and it is the function of quantum theory to regularize, interpret, and make predictions about them. Specifically, quantum physics deals with problems of this kind: Given an actual (in contradistinction to a *Gedanken*-type) repeatable laboratory procedure Π for the *preparation* of physical systems, what will be the statistical distribution of measurement results obtained from observations performed upon an ensemble of systems all prepared in accordance with Π? This question may refer to any observable and to measurements at any given time after preparation.

Classically, measurement results are simply revelations of the values of observable properties *possessed* by the system. The key word here is *possessed,* for it expresses succinctly the classical and indeed the common-sense relationship between measurement results and observables. In quantum mechanics the connection is a weaker one. It is no longer possible to pictorialize physical systems as objects characterizable by definite values of the observables. The possessive adherence of observables to systems fails. This peculiarity of quantum observables has been characterized by one of the authors as *latency.*[10] A brief explanation of the idea of latency relevant to the present problem is given in Ref. 9.

The possessed quality of all classical observables brought the ideas of measurement and preparation conceptually close to one another. Since a measurement operation simply revealed a possessed value, the same operation could also be called a preparation method for obtaining systems having that value of the measured observable. In quantum theory, however, the constructs measurement and preparation must be disjoined. Failure to do so leads to an erroneous interpretation of the projection postulate.[11] The correlation between measured values and the state of a system is less direct, and thus the idea of measurement needs more careful analysis than is ordinarily necessary.

To measure the *position* of an object, one juxtaposes the object with a scale and identifies its position with the scale mark. No further analysis is required, because the tacit belief that the object *has* the position avoids every complication. In this simple sense, a position measurement is the establishment of a self-evident correspondence between a set of numbers and the values of an observable—self-explanatory and self-validating. This manner of correspondence between measured numbers and values of the observable is part of every measurement in classical physics as well as in quantum mechanics. It is often called "direct" measurement, and we shall designate it by $\mathscr{M}_1$.*

The measurement of *velocity* is a little less obvious. If the speed of a moving vehicle is to be determined, the reading of a speedometer can be noted. What this measurement delivers directly via the $\mathscr{M}_1$ concept is the position of the needle. To interpret this position as a velocity requires assumptions of an interpretative, theoretical sort, ideas beyond the content of the mere association $\mathscr{M}_1$. It involves a mathematical analysis of the instrument, which leads to proof of a further correlation between the possessed position values of the

* In this essay we employ script letters ($\mathscr{M}$, $\mathscr{A}$, $\mathscr{B}$, . . .) to designate observables [the general concept of measurement is regarded as an observable]; capital italic letters (A, B) denote operators and lowercase italic letters (a, b) numerical measured values; lowercase Greek letters (α, β), except ρ (the density operator), denote quantum state vectors.

needle and the (in this case possessed) speed values of the moving vehicle. The ingredient of the measurement concept which establishes this theoretical correlation will be denoted by $\mathscr{M}_2$. Thus $\mathscr{M}_2(\mathscr{A}, \mathscr{B}, \ldots)$ represents any empirical procedure yielding numbers $a, b, \ldots$ which *through a theory* can be interpreted as the values *associated* with observables $\mathscr{A}, \mathscr{B}, \ldots$.

The fundamental difference between classical and quantal physics lies in their respective conceptions of the nature of this *association* between the measured values of observables and the systems with which the observables are said to be associated. Consider again the speedometer. Classically, its operation was explained by an $\mathscr{M}_2$ theory, which established a correlation between the *possessed* position values of the needle and the *possessed* speed values of the vehicle. Quantum mechanically, a correlation cannot be so simply described, because the intrinsic *latency* of quantum observables disavows the classically implicit presupposition that the system *possesses* any speed value to be correlated with the needle position. What construct then plays the same role in quantum physics that possession does in classical physics?

An examination of quantum axiomatics reveals (cf. Ref. 3) that the strongest kind of correlation statement for the speedometer which can be conceived within the framework of quantum physics has this form: The joint probability that a *measurement* of the needle position will yield x_k and a simultaneous *measurement* of the speed of the vehicle will yield v_l vanishes unless $l = k$.

Thus in quantum mechanics the classical notion of *possession* is replaced by a primitive construct, commonly called *measurement*, which is not, however, the concept of measurement designated above as $\mathscr{M}_2$. Indeed, as just exemplified, an $\mathscr{M}_2$ cannot be described quantum theoretically, except in terms of this primitive measurement construct, which will henceforth be called $\mathscr{M}_1$ to indicate that it is an abstract generalization of the elementary act of "direct" observation which was earlier denoted by $\mathscr{M}_1$. Since quantum physics, owing to the characteristic latency of its observables, cannot make statements about correlations among possessed values of observables, it speaks instead of correlations among the results of primitive "direct" measurements $\mathscr{M}_1$, which are in practice *theoretical* constructs which can no more be performed "directly" than could the possessed values of, say, the speed of a classical body be perceived "directly."

For a microphysical illustration, consider the measurement of electron position using a scintillation screen. The theory ($\mathscr{M}_2$) which justifies identification of the location of the scintillation with the position of an electron establishes a correlation not between two possessed attributes (positions of electron and scintillation) but between two $\mathscr{M}_1$'s: one the genuinely "direct" observation of the scintillation, the other the more sophisticated, abstract, and theoretically primitive relation that the observable position bears to an electron in quantum physics. These $\mathscr{M}_1$'s are governed by certain basic axioms to which Sec. 2 will be devoted.

Using the foregoing concepts, it is possible to define precisely what is meant by simultaneous measurability of two observables: *Observables $\mathscr{A}$ and $\mathscr{B}$ will be termed compatible, simultaneously measurable, or jointly measurable if there exists an $\mathscr{M}_2(\mathscr{A}, \mathscr{B})$*, that is, an operation furnishing two numbers a, b with the same probabilities that quantum theory confers upon the two propositions "$\mathscr{M}_1(\mathscr{A})$ yields a" and "$\mathscr{M}_1(\mathscr{B})$ yields b," where both $\mathscr{M}_1$'s refer to the same instant in time. The *compatibility problem*, to which the rest of this paper is devoted, may therefore be stated as follows: If $\mathscr{A}$, $\mathscr{B}$ are noncommuting observables, is it quantum theoretically possible for an $\mathscr{M}_2(\mathscr{A}, \mathscr{B})$ to exist?

2. Quantum Axiomatics and the Uncertainty Theorem

The basic postulates of quantum physics will now be stated, and several important theorems will then be reviewed.

P1 CORRESPONDENCE POSTULATE

(Some) linear Hermitian operators on Hilbert space which have complete orthonormal sets of eigenvectors correspond to physical observables. If operator A corresponds to observable $\mathscr{A}$, then the operation $\mathscr{F}(A)$, where $\mathscr{F}$ is a function, corresponds to observable $\mathscr{F}(\mathscr{A})$.

We shall use the symbol $\leftrightarrow$ to represent this operator-observable correspondence; thus $A \leftrightarrow \mathscr{A}$ means A "corresponds to" $\mathscr{A}$ in the sense of P1. The observable $\mathscr{F}(\mathscr{A})$ is defined operationally as follows: Measure $\mathscr{A}$ and use the result a to evaluate the given function $\mathscr{F}$; the number $\mathscr{F}(a)$ is then the result of an $\mathscr{F}(\mathscr{A})$-measurement. The function $\mathscr{F}$ of operator A, $\mathscr{F}(A)$, is found by the following standard mathematical procedure: Consider the spectral expansion of A,

$$A = \sum_k a_k P_{\alpha_k},$$

where a_k is an eigenvalue and P_{α_k} denotes the projector onto the span of eigenvector α_k; the operator $\mathscr{F}(A)$ is then simply

$$\sum_k \mathscr{F}(a_k) P_{\alpha_k}.$$

P2 MEAN VALUE POSTULATE

To every ensemble of identically prepared systems there corresponds a real linear functional of the Hermitian operators, $m(A)$, such that if $A \leftrightarrow \mathscr{A}$, the value of $m(A)$ is the arithmetic mean $\langle \mathscr{A} \rangle$ of the results of $\mathscr{A}$-measurements* performed on the member systems of the ensemble.

The content of P1 and P2 is slightly different from von Neumann's axiomatizations. In the original form of the Correspondence Postulate, observables and Hermitian operators were assumed to stand in one-to-one correspondence; the postulate included both of the following statements:

(1) Every observable has an Hermitian operator representative.

(2) Every Hermitian operator corresponds to a physical observable.

* Reference is here to the primary quantum-measurement construct $\mathscr{M}_1(\mathscr{A})$.

In 1952, Wick, Wightman, and Wigner[12] effectively challenged the symmetry of this quantal correspondence by introducing the concept of superselection rules, i.e. assertions which declare certain Hermitian operators to be unobservable in principle. To embrace superselection rules with minimal theoretic change, the word *every* in (2) is replaced by *some*:

(2′) Some Hermitian operators correspond to physical observables.

Just as superselection rules challenge the word *every* in (2), an important facet of the compatibility problem hinges on the word *every* in proposition (1). Accordingly, the need will arise subsequently to distinguish between different "degrees" of operator-observable correspondence. For this purpose the following terminology will be adopted: *Strong* correspondence means that both (2′) *and* (1) are assumed; *weak* correspondence means that the Correspondence Postulate includes (2′) but *not* (1), as in P1.

In subsequent sections, the relationship of this choice of correspondence schemes to the problem of compatibility will be developed, and we shall demonstrate that only the weak type (P1) is acceptable. Henceforth we distinguish between P1S—von Neumann, strong—and our P1.

Several "elementary" quantum theorems will now be stated without proof. Although the content of these theorems is well known, the fact is not always acknowledged that they are *theorems*, i.e. derivable from P1 and P2: P1 and P2 (or their equivalents) rigorously *imply* all the general propositions of quantum statics.

TH. 1[13]

For each mean value functional $m(A)$ there exists an Hermitian operator ρ such that for each A,

$$m(A) = \mathrm{Tr}\,(\rho A).$$

The Hermitian operator ρ, known as the statistical operator or density operator, is not only an "index" of measurement statistics but also the seat of causality in quantum physics. For this reason, we shall call ρ the quantum *state* of the ensemble to which it refers. The general "law of motion" is given by the following axiom.

P3 DYNAMICAL POSTULATE

To every type of closed quantum system there corresponds a linear unitary operator T (the evolution operator) such that the temporal development of the density operator ρ for an ensemble of like systems is given by

$$\rho(t_2) = T(t_2, t_1)\rho(t_1)T^{\dagger}(t_2, t_1).$$

In the following theorems, we assume the Hermitian operators to have discrete spectra; but similar propositions hold for the continuous case too.

TH. 2

The probability $W_{\mathscr{A}}(a_k; \rho)$ that an $\mathscr{A}$-measurement on a system from an ensemble with density operator ρ will yield the A-eigenvalue a_k is given by

$$W_{\mathscr{A}}(a_k; \rho) = \mathrm{Tr}\ (\rho P_{\mathscr{H}_k}),$$

where $\mathscr{H}_k$ is the subspace belonging to a_k.

TH. 3

$\mathrm{Tr}\ \rho = 1.$

TH. 4

The only possible results of $\mathscr{A}$-measurements are the eigenvalues $\{a_k\}$ of A, where $A \leftrightarrow \mathscr{A}$.

TH. 5

The density operator ρ is positive semidefinite.

Careful analysis (cf. Ref. 9) of the foregoing theorems reveals that only *weak* correspondence need be invoked to prove them; they would of course still follow, however, if P1 were replaced by an axiom of strong correspondence, namely

P1S
The set of physical observables is in one-to-one correspondence with the set of linear Hermitian operators on Hilbert space with complete orthonormal sets of eigenvectors. Thus, if $A \leftrightarrow \mathscr{A}$, then $\mathscr{F}(A) \leftrightarrow \mathscr{F}(\mathscr{A})$.

A cursory examination of P1S and P2 seems to suggest that nothing about simultaneous measurement could ever be derived from such axioms, for in them reference is made only to measurements of single observables via $\mathscr{M}_1(\mathscr{A})$. Indeed, the absence of a similar joint measurement construct $\mathscr{M}_1(\mathscr{A}, \mathscr{B}, \ldots)$ appears to justify the conclusion that quantum theory is noncommittal to the problem of compatibility and that, in order to discuss simultaneous measurements at all, P2 must be augmented by some kind of joint probability postulate. This indifference is illusory, for P1S and P2 do in fact place severe restrictions upon simultaneous measurements through a theorem to be reviewed in Sec. 4.

To approach the problem of joint measurements from an axiom set referring only to single measurements, it is necessary to develop a theory of *compound* observables, i.e. observables defined as functions of several ordinary observables. Then information regarding joint measurements can be extracted from an analysis of single measurements defined as functions of the joint measurement results. For example, a compound observable $\mathscr{F}(\mathscr{A}, \mathscr{B})$ may be operationally defined as follows: Measure $\mathscr{A}$ and $\mathscr{B}$ simultaneously, and substitute the results a and b into the function $\mathscr{F}(a, b)$; the value $f = \mathscr{F}(a, b)$ is then the result of the $\mathscr{F}(\mathscr{A}, \mathscr{B})$-measurement. Then by P1S, there exists an operator F to represent $\mathscr{F}(\mathscr{A}, \mathscr{B})$; hence if F is known, $\mathscr{F}(\mathscr{A}, \mathscr{B})$-measurements are subject to quantum-mechanical analysis, and in this sense joint measurements would be in the domain of the ordinary quantum theory of $\mathscr{M}_1$'s.

This leads to an old quantum problem.[14] Given the correspondences $A \leftrightarrow \mathscr{A}, B \leftrightarrow \mathscr{B}, \ldots$ and a compound observable $\mathscr{F}(\mathscr{A}, \mathscr{B}, \ldots)$, what F corresponds to $\mathscr{F}$? If P1S is adopted, the existence of such an F is assured (provided, of course, that $\mathscr{A}, \mathscr{B}, \ldots$ are simultaneously measurable). If, however, only the weaker P1 holds, the existence of an F such that $F \leftrightarrow \mathscr{F}(\mathscr{A}, \mathscr{B}, \ldots)$ is by no means guaranteed. In neither case is there a general prescription for finding F, to be sure; but it is obviously necessary that all deductions based on a proposed F be consistent with P2, the definition of $\mathscr{F}$, and the theorems reviewed above. Th. 1 and Th. 4 particularly suggest useful consistency conditions. To formulate them, we employ the following notation.

Define the sets $\mathscr{E}(A)$ and $\mathscr{N}(\mathscr{F})$ thus: $\mathscr{E}(A)$ comprises the eigenvalues belonging to the operator A; $\mathscr{N}(\mathscr{F})$ is the set of obtainable measurement results associated with an observable $\mathscr{F}$. When $\mathscr{F} = \mathscr{A}$, then $\mathscr{N}(\mathscr{A}) = \mathscr{E}(A)$ by Th. 4. However, when $\mathscr{F}$ is a function, let us say of $\mathscr{A}$ and $\mathscr{B}$, then it is possible that correlations between $\mathscr{A}$ and $\mathscr{B}$ might preclude the occurrence of certain a priori conceivable values of $\mathscr{F}$, i.e. preclude certain of the values $\mathscr{F}(a_k, b_l)$ calculable from eigenvalues of A and B under the a priori assumption that all eigenvalue pairs (a_k, b_l) are possible. In this contingency, $\mathscr{E}(F) \subset \mathscr{N}(\mathscr{F})$. Finally, for a state ρ, let $W(a_k, b_l, \ldots; \rho)$ denote the joint probability that simultaneous $\mathscr{A}, \mathscr{B}, \ldots$-measurements yield $a_k, b_l, \ldots$.

Two consistency conditions may then be expressed in this manner: If $F \leftrightarrow \mathscr{F}(\mathscr{A}, \mathscr{B}, \ldots)$, then

$$(\mathrm{C}_1) \quad \sum_{kl} W(a_k, b_l, \ldots; \rho)\mathscr{F}(a_k, b_l, \ldots) = \mathrm{Tr}\,(\rho F),$$

for every ρ, and

$$(\mathrm{C}_2) \quad \mathscr{E}(F) \subseteq \mathscr{N}[\mathscr{F}(\mathscr{A}, \mathscr{B}, \ldots)].$$

Condition (C_1) arises from Th. 1 and the definition of $\mathscr{F}$, while (C_2) is needed to prevent conflict with Th. 4. The usefulness of (C_1) must however be questioned, for the joint probability W is so far unknown. Nevertheless, this condition is not undiscriminating, since for the proper choice of $\mathscr{F}$ it becomes independent of the form of W. (Cf. Ref. 9.)

Both P1 and P1S involve explicit postulation of the correspondence $\mathscr{F}(A) \leftrightarrow \mathscr{F}(\mathscr{A})$, and a survey[15] of the proofs of Theorems 1–5 indicates clearly the value of that rule. Since Th. 2 (i.e. the form of $W_{\mathscr{A}}$) is the cornerstone of practical calculations in quantum theory and therefore not a proposition that could easily be challenged, the following theorem suggests very strongly that the correspondence $\mathscr{F}(A) \leftrightarrow \mathscr{F}(\mathscr{A})$ could not reasonably be removed from the quantum axiom set.

CONSISTENCY THEOREM

If $W_{\mathscr{A}}(a_k; \rho) = \mathrm{Tr}\,(\rho P_{\mathscr{H}_k})$ and if there exists an operator F such that $F \leftrightarrow \mathscr{F}(\mathscr{A})$, then $F = \mathscr{F}(A)$, where $A \leftrightarrow \mathscr{A}$. The proof is simple.

The operator F must satisfy consistency condition (C_1):

$$\sum_k \mathrm{Tr}\,(\rho P_{\mathscr{H}_k})\mathscr{F}(a_k) = \mathrm{Tr}\,(\rho F).$$

Thus,

$$\mathrm{Tr}\,[\rho(F - \sum_k \mathscr{F}(a_k)P_{\mathscr{H}_k})] = 0 \qquad \text{for every } \rho,$$

which implies

$$F = \sum_k \mathscr{F}(a_k)P_{\mathscr{H}_k} = \mathscr{F}(A).$$

This result also satisfies (C_2).

The special case of $\mathscr{F}(\mathscr{A})$ that has received most attention is a fairly complicated one: $\mathscr{F}(\mathscr{A}) = (\mathscr{A} - \langle\mathscr{A}\rangle)^2$, where $\langle\mathscr{A}\rangle$ is a real constant which is the arithmetic mean of the $\mathscr{M}_1(\mathscr{A})$'s on the ensemble of interest. Using P2, we see that $\mathscr{F}(\mathscr{A}) = (\mathscr{A} - m(A))^2$; then, by the correspondence rule in P1, $\mathscr{F}(\mathscr{A}) \leftrightarrow (A - m(A)1)^2$. By definition,

$$(\Delta\mathscr{A})^2 = m[(A - m(A)1)^2] = \langle(\mathscr{A} - \langle\mathscr{A}\rangle)^2\rangle;$$

the standard deviation $\Delta\mathscr{A}$ is a common statistical quantity defined as a function of measurement results from an ensemble.

Often, $\Delta\mathscr{A}$ has been linked—erroneously, we believe—to the problem of compatibility by way of the Heisenberg uncertainty principle. A few remarks seem appropriate here in order to invalidate the popular contention that the uncertainty principle places restrictions on simultaneous measurability. Heisenberg's principle is a theorem, rigorously derivable from the quantum postulates; it states that under fairly general conditions,

$$\Delta\mathscr{A}\ \Delta\mathscr{B} \geq \tfrac{1}{2}|m([A, B])|,$$

where A, B are Hermitian operators representing quantum observables $\mathscr{A}, \mathscr{B}$, and $\Delta\mathscr{A}, \Delta\mathscr{B}$ refer to collectives of $\mathscr{A}$- and $\mathscr{B}$-measurements.

The principal point to be stressed here is that $\Delta\mathscr{A}$ and $\Delta\mathscr{B}$ have physical meaning only within the context of *statistics*. It is therefore illogical to interpret the uncertainty principle as a denial of the possibility of simultaneous measurement of $\mathscr{A}$ and $\mathscr{B}$ upon a single system if $[A, B] \neq 0$, as has sometimes been done. The only sense in which $\Delta\mathscr{A}\ \Delta\mathscr{B}$ may refer to a single system is purely *statistical*, i.e. to an ensemble involving *one* system sequentially measured and reprepared. Furthermore it should be noted that the product $\Delta\mathscr{A}\ \Delta\mathscr{B}$ is not even calculated from *simultaneous* measurements of $\mathscr{A}$ and $\mathscr{B}$ performed on each system. Thus, whatever conclusions one may reach concerning the notion of compatibility, i.e. simultaneous measurability of several observables on a single system, there can be no conflict with the uncertainty principle, a relation involving statistical properties of measurements of single observables.

To summarize: Regardless which propositions about joint measurements may or may not be consistently incorporated into quantum theory, the uncertainty principle remains unscathed so long as its interpretation does not transcend the content justified by its proof. Conversely, the uncertainty principle is not an a priori restriction on any consideration purely about joint measurements.

Strictly construed, the uncertainty principle is irrelevant to the problem of compatibility.

3. Trivial Joint Measurements and Commutability

There is a type of joint measurement whose consistency with quantum theory is never questioned, for it involves the performance of only one measurement upon the system. The resulting number is then used to generate a set of numbers through a set of established functions; the simultaneous measurement of a set of observables has thus been performed, albeit in a rather trivial sense. Such joint measurements, performed simply by arithmetical manipulation of one measurement result for a single observable, will henceforth be called *trivial joint measurements.*

The question then arises as to whether the joint measurement of any two observables is reducible to a trivial joint measurement; if so, quantum theory could embrace the concept of simultaneous measurement in a very natural way. However, the correspondence rule $\mathscr{F}(A) \leftrightarrow \mathscr{F}(\mathscr{A})$ may be used to prove that any two operators jointly measurable in this trivial sense necessarily commute.

These considerations do not imply that noncommuting observables are incompatible; they merely establish that such observables are not trivially compatible. Nevertheless, since $[A, B] = 0$ is (1) a necessary condition for trivial joint measurability of $\mathscr{A}$ and $\mathscr{B}$ and (2) the only condition under which $\Delta\mathscr{A}\,\Delta\mathscr{B} = 0$ may hold, it is sometimes claimed (via one of the misinterpretations of the uncertainty principle) that the only simultaneous measurements permitted by quantum theory are the trivial ones, that commutability is the mathematical criterion of compatibility. But in view of our preceding remarks about the uncertainty principle, such a position is evidently not tenable.

Although the notion of trivial joint measurement is not an adequate basis for a general treatment of simultaneous measurements, it does provide a means for deriving the joint probabilities associated with several *commuting* observables.

Omitting the detailed proof,[16] we state here only the essential result. Let $\mathscr{A}$ and $\mathscr{B}$ be simultaneously measurable through an auxiliary variable $\mathscr{C}$ defined by

$$\mathscr{A} = \mathscr{F}(\mathscr{C}), \mathscr{B} = \mathscr{G}(\mathscr{C}),$$

supposing that $[A, B] = 0$.

The joint probability that $\mathscr{M}_1(\mathscr{A})$ and $\mathscr{M}_1(\mathscr{B})$ will yield (a_k, b_l) for the state ρ is then given uniquely by

$$W(a_k, b_l; \rho) = W_{\mathscr{C}}(c_{kl}; \rho) = \mathrm{Tr}\ (\rho P_{\gamma_{kl}}),$$

where the new symbols are defined through the spectral expansion of C,

$$C = \sum_{kl} c_{kl} P_{\gamma_{kl}}.$$

This analysis leads[17] to a proposition of some importance, namely: (J) The joint probability $W(a_k, b_l; \rho)$, $[A, B] = 0$, is a unique functional of the state ρ; thus the *state* of an ensemble is sufficient to determine the distribution, as would be the case in classical physics. In particular, no additional information regarding the method of measurement is needed to obtain W; once a W for a given ρ is found by the method of trivial joint measurement, it may be assumed that it is *the* W associated with the given ρ, independently of how $\mathscr{A}$ and $\mathscr{B}$ might be measured.

Suppose, however, that $[A, B] \neq 0$. Then the method of trivial joint measurements is of course inapplicable. Does (J) still hold? Is the quantum state ρ alone sufficient to determine W's for joint measurements of *noncommuting* observables? We shall study this matter in later sections.

4. Von Neumann's Theorem: Noncommuting Observables Are "Incompatible"

The popular belief that the *only* compatible observables are the trivially compatible ones was reviewed in Sec. 2, where the uncertainty principle, the standard basis of this dogma, was presented and found irrelevant. However, there exists also a rather formidable logical demonstration that if two observables are compatible they are trivially compatible. It is an elegant theorem,[18] due to von Neumann, that strangely enough appears to be almost universally ignored, even by proponents of the viewpoint for which it is the strongest support. Indeed the main impact of the theorem seems to have been to influence mathematicians[19] interested in modern physics to *define* the term "simultaneously measurable" by the commutability condition for trivial joint measurability, which is not very helpful in view of the fact that both words in common physical usage already had other definitions, as explained in Sec. 1. Because von Neumann's theorem is of central importance to the problem of compatibility, it is appropriate here to consider it, even though space does not permit a full analysis of the hypotheses on which it is based. It reads:

SIMULTANEOUS MEASURABILITY THEOREM

If $\mathscr{A}$ and $\mathscr{B}$ are compatible and $\mathscr{A} \leftrightarrow A$, $\mathscr{B} \leftrightarrow B$, then $[A, B] \doteq 0$.

Expressed succinctly, the theorem says that if $\mathscr{A}$ and $\mathscr{B}$ are compatible, they are trivially compatible, for their operators necessarily commute. Unlike the semiclassical *Gedankenexperimente*, the vague interpretations of the uncertainty principle, and some strange philosophizing about subjective wave-packet reductions, the theorem offers an argument strong and clear in behalf of the proposition that noncommuting observables cannot even in principle be measured simultaneously. It affirms that the very notion of general compatibility simply cannot logically be appended to the established theoretical structure of quantum physics, *unless* the latter is somehow modified. This possibility of nullifying the theorem by such a basic alteration in the quantum postulates will be considered later.

We note here merely that von Neumann's proof crucially involves the postulate of strong correspondence, discussed in Sec. 2.

Inspired by the preceding theorem, various authors[20] have suggested that quantum mechanics should be rephrased in a new *logical* framework which would properly allow for incompatibility. We believe, and intend to show in subsequent sections of this article, that von Neumann's mathematics does not in fact establish incompatibility as an intrinsic quantal property. Hence, if our analysis is correct, any "quantum logic" designed to embrace incompatibility is founded upon a mistaken interpretation of quantum physics. We shall now expose certain salient features of so-called "quantum logic," in order to establish its relation to von Neumann's theorem.

Propositions, or questions, can be introduced into quantum theory as functions of observables. Consider an observable $\mathscr{A} \leftrightarrow A = \sum_k a_k P_{\alpha k}$ and the proposition $\mathscr{P}_n$: "$\mathscr{M}_1(\mathscr{A})$ will yield a_n." Proposition $\mathscr{P}_n$ is simply the observable "measured" as follows: Measure $\mathscr{A}$; if a_n results, assign to $\mathscr{P}_n$ the value 1; if $a_k (\neq a_n)$ emerges, assign to $\mathscr{P}_n$ the value 0. In short, $\mathscr{P}_n = \mathscr{F}_n(\mathscr{A})$, where $\mathscr{F}_n$ is defined by $\mathscr{F}_n(a_k) = \delta_{nk}$. Hence

$$\mathscr{P}_n \leftrightarrow \mathscr{F}_n(A) = \sum_k \mathscr{F}_n(a_k) P_{\alpha k} = P_{\alpha n}.$$

A suitable projection operator may also be found for any proposition involving commuting observables; but because of von Neumann's theorem, any compound proposition involving non-commuting observables must be regarded as undecidable, or absurd. For any two compatible propositions $\mathscr{P}$ and $\mathscr{Q}$, it is possible to find operators corresponding to the logical relation $\mathscr{P}$ "or" $\mathscr{Q} \equiv \mathscr{P} \cup \mathscr{Q}$ and $\mathscr{P}$ "and" $\mathscr{Q} \equiv \mathscr{P} \cap \mathscr{Q}$:

$$\mathscr{P} \cup \mathscr{Q} \leftrightarrow P + Q - PQ,$$

$$\mathscr{P} \cap \mathscr{Q} \leftrightarrow PQ.$$

The change in logic said to be necessitated by quantum mechanics has to do with the classical distributive law of propositions:

$$\mathscr{P} \cap (\mathscr{Q} \cup \mathscr{R}) = (\mathscr{P} \cap \mathscr{Q}) \cup (\mathscr{P} \cap \mathscr{R}).$$

Suppose, for example, that A, B are operators in a two-dimensional Hilbert space. If $[A, B] \neq 0$, and $\mathscr{P} \leftrightarrow P_{\beta_1}$, $\mathscr{Q} \leftrightarrow P_{\alpha_1}$, $\mathscr{R} \leftrightarrow P_{\alpha_2}$, then, because of von Neumann's theorem, the distributive law cannot hold in quantum theory. This stems from the correspondence

$$\mathscr{P} \cap (\mathscr{Q} \cup \mathscr{R}) \leftrightarrow P_{\beta_1}(P_{\alpha_1} + P_{\alpha_2} - P_{\alpha_1}P_{\alpha_2}) = P_{\beta_1}(1 - 0) = P_{\beta_1};$$

but $(\mathscr{P} \cap \mathscr{Q}) \cup (\mathscr{P} \cap \mathscr{R})$ is an absurd proposition, for neither $\mathscr{P} \cap \mathscr{Q}$ nor $\mathscr{P} \cap \mathscr{R}$ is measurable, since they are compounds of $\mathscr{A}$ and $\mathscr{B}$ with $[A, B] \neq 0$. Thus, since the distributive law apparently cannot hold in quantum theory, it has been suggested that some "nondistributive" logic is required for quantum propositions.

We shall return to this point in Sec. 7.

5. Counterexamples Suggesting That Noncommuting Observables Are Compatible

Mathematically, von Neumann's simultaneous measurability theorem is beyond criticism; it is a legitimate deduction from P1S and P2. If, therefore, one could find a counterexample, i.e. describe *quantum mechanically* a physical process fully certifiable as a simultaneous measurement of, say, position and momentum, then the basis of von Neumann's theorem would require reformulation. It would then establish not the incompatibility of physical observables but rather the *inconsistency of the quantum-mechanical axioms.* It is possible to construct such counterexamples, and two of them will be recorded.

Consider first the quantum theory connected with the measurement of a single observable, viz. the "time-of-flight" method for measuring the momentum $\mathscr{P}$ of an electron. The rule of correspondence for position $\mathscr{X}$ might consist, for example, of the direct observation of a coincidence between a scale mark and a macroscopic spot appearing on a photographic plate in response to an electron impact.

An "electron gun" prepares the state $\rho = P_\psi$. Using non-relativistic wave mechanics, we find that the probability density $w_{\mathscr{P}}(p; \psi)$ for $\mathscr{M}_1(\mathscr{P})$ at the time of preparation is

$$w_{\mathscr{P}}(p; \psi) = \left| \frac{1}{\sqrt{2\pi\hbar}} \int_{-\infty}^{\infty} e^{-ipx/\hbar} \psi(x)\, dx \right|^2 .$$

This distribution is the quantum-mechanical test for deciding whether a proposed experiment which generates numbers via the established operational definition for $\mathscr{M}_1(\mathscr{X})$ qualifies as a momentum measurement scheme $\mathscr{M}_2(\mathscr{P})$. If the numbers in question are to be regarded as $\mathscr{M}_1(\mathscr{P})$-results, they must satisfy the theoretical distribution $w_{\mathscr{P}}(p; \psi)$.

Let $t = 0$ be the time when the electron is known to be in the prepared state $\rho = P_\psi$. The wave function $\psi(x, t = 0)$ is assumed to be of compact support, and it is convenient to set up the origin of the x-axis so that the interval where $\psi(x) \not\equiv 0$ is $(-x_0, x_0)$. The $\mathscr{M}_2(\mathscr{P})$-procedure[21] is simple: We simply wait a very long time $(t \to \infty)$ as the electron moves *freely*, and then we measure the observable $\mathscr{F}(\mathscr{X}) = m\mathscr{X}/t$, where m is the electron mass. The number obtained then counts as the result of $\mathscr{M}_1(\mathscr{P})$ at $t = 0$. To justify this operational definition of $\mathscr{P}$ quantum mechanically, one must prove that the probability for $\mathscr{M}_1(\mathscr{P})$ to yield $p \in (p_1, p_2)$ at $t = 0$ equals the probability that $\mathscr{M}_1[\mathscr{F}(\mathscr{X})]$ yields $(m\mathscr{X}/t) \in (p_1, p_2)$ at $t \to \infty$. The details are given in Ref. 9.

We conclude that the results of "direct" $\mathscr{F}(\mathscr{X})$-measurements, performed sufficiently long after the preparation of $\psi(x, 0)$, will be distributed just like the theoretical results for $\mathscr{M}_1(\mathscr{P})$ upon $\psi(x, 0)$. This time-of-flight arrangement is therefore fully certified quantum mechanically as an operational definition of $\mathscr{P}$. Because quantum theory can make only statistical predictions, no further guarantee that this method "really" makes $\mathscr{P}$-measurements is required. Indeed, further *quantal* analysis of the question is theoretically meaningless.

One can also see that this time-of-flight method for obtaining the results which $\mathcal{M}_1(\mathcal{P})$ at $t = 0$ would yield, determines likewise the result which $\mathcal{M}_1(\mathcal{P})$ would yield any time $t > 0$. This follows at once from the fact that momentum is conserved in the free motion of the electron; in quantum-mechanical terms,

$$W_{\mathcal{P}}[p \in (p_1, p_2); \psi(x, 0)] = W_{\mathcal{P}}[p \in (p_1, p_2); \psi(x, t)].$$

Hence, by the same reasoning which validated the time-of-flight method as a rule of correspondence for $\mathcal{M}_1(\mathcal{P})$ at $t = 0$, we can regard the results of $\mathcal{M}_1[\mathcal{F}(\mathcal{X})]$, $t \to \infty$, as $\mathcal{M}_1(\mathcal{P})$-results for *any* $t > 0$. In particular, consider the instant when the electron strikes the photographic plate and the result emerges. For that instant we may conclude with full quantum-mechanical justification that $\mathcal{M}_1(\mathcal{P})$ would have yielded $\mathcal{F}(x)$ where x is the result of the $\mathcal{M}_1(\mathcal{X})$. Contrary to the prohibitions of von Neumann's theorem, we have here an empirical method for the simultaneous measurement of $\mathcal{X}$ *and* $\mathcal{P}$, two renowned noncommuting observables!

There is a tendency to dismiss simultaneous measurement schemes such as the one just described as if they did not in fact *legitimately* challenge the orthodox view. One authoritative argument was first employed by Heisenberg and may be summarized by his statement[22] that "the uncertainty relation does not refer to the past." In the time-of-flight experiment, by the time the $\mathcal{X}$, $\mathcal{P}$-values emerge, the time to which they refer—the instant just prior to the electron's collision with the photographic plate—is past; and the electron is then buried in the plate. According to Heisenberg, such "knowledge of the past is of a purely speculative character, since it can never . . . be used as an initial condition in any calculation of the future progress of the electron and thus cannot be subjected to experimental verification. It is a matter of personal belief whether such a calculation concerning the past history of the electron can be ascribed any physical reality or not."[23]

In rejoinder to this distinctly philosophical and somewhat subjective disposal of the matter, we offer the following comments.

The word *knowledge* is not unambiguous when employed in discussions regarding quantum measurement. As we have seen, from a strict quantal point of view, an electron *never possesses properties* $\mathscr{X}$, $\mathscr{P}$ of which one can be knowledgeable or ignorant. (There does not exist a preparation scheme Π which produces electrons always yielding the same measured $\mathscr{X}$, $\mathscr{P}$-values.) Accordingly, measurement should never be described as though it increased knowledge by revealing perhaps with growing precision the actual, previously unknown, "value" of an observable. Measurements merely generate numerical results associated with certain operations upon the system of interest. The meaning of these numbers is provided by the theory into which they are fed; in quantum theory the numbers are not to be regarded as measures of possessed attributes.

It is therefore not very meaningful to say that the uncertainty relations do not refer to the past. They refer to the standard deviations of collectives of measurement results at any time and have no bearing on measurements upon a single system at a single time, since standard deviations refer only to measurements upon ensembles. Hence, as already explained in Sec. 2, the emergence of simultaneous $\mathscr{X}$, $\mathscr{P}$-values upon measurement in no way violates the uncertainty principle.

In the time-of-flight method, the $\mathscr{X}$, $\mathscr{P}$-measurement results-admittedly refer to the instant just prior to the electron's impact on the plate. They are indeed useless for predicting (in classical fashion) the result of a future $\mathscr{X}$-measurement, yet they are no more "speculative" or lacking in "physical reality" than any other measurement result. Their lack of predictive power stems from the fact that the "motion" of quantum systems is not governed by Newtonian laws. Reference of the $\mathscr{X}$, $\mathscr{P}$-values to a past time is no special feature of *simultaneous* measurements; it is characteristic of all quantum measurements. Certainly, the time-of-flight measurement of $\mathscr{P}$ alone referred to $t = 0$, although the result did not emerge until $t \rightarrow \infty$. Nevertheless, such $\mathscr{P}$-measurements play a key role in the process of empirical verification; for example, their statistical distribution determines whether or not the state prepared by the "electron gun" is really $\psi(x, 0)$. Surely, if the physical significance of such $\mathscr{X}$, $\mathscr{P}$-values is a matter of "personal belief," then all measurement results for *single* observables are likewise merely of solipsistic significance.

We are therefore forced to conclude that the foregoing method for simultaneous measurement of $\mathscr{X}, \mathscr{P}$ is as significant as any other quantum-mechanical measurement scheme.

We now present another counterexample to the simultaneous measurability theorem. Consider two quantum systems S_1, S_2 with observables $\mathscr{A}_1$, $\mathscr{B}_1$, and $\mathscr{A}_2$ associated with S_1 and S_2, respectively. Let $[A_1, B_1] \neq 0$ and denote eigenvectors and eigenvalues as follows:

$$A_1\alpha_k^{(1)} = a_k^{(1)}\alpha_k^{(1)}, \qquad A_2\alpha_k^{(2)} = a_k^{(2)}\alpha_k^{(2)}.$$

Although S_1 and S_2 are noninteracting, they are assumed to be in a correlated state:

$$\Psi = \sum_k c_k\alpha_k^{(1)} \otimes \alpha_k^{(2)}.$$

If $\mathscr{A}_2$ has an operational definition, the correlation in Ψ that relates $\mathscr{M}_1(\mathscr{A}_1)$-results to $\mathscr{M}_1(\mathscr{A}_2)$-results may be used to construct an $\mathscr{M}_2(\mathscr{A}_1)$. As in the $\mathscr{M}_2(\mathscr{P})$ case of our previous example, we must establish a theoretical matching between probabilities associated with $\mathscr{M}_1(\mathscr{A}_2)$ and $\mathscr{M}_1(\mathscr{A}_1)$. Since

$$[A_1, A_2] = [A_1 \otimes 1, 1 \otimes A_2] = 0,$$

$\mathscr{A}_1$ and $\mathscr{A}_2$ may be jointly measured (trivially) through an auxiliary observable (cf. Sec. 3). The joint probability $W(a_k^{(1)}, a_l^{(2)}; \Psi)$ is then

$$W(a_k^{(1)}, a_l^{(2)}; \Psi) = \mathrm{Tr}\,(P_\Psi P_{\alpha_k^{(1)} \otimes \alpha_l^{(2)}}) = |c_k|^2\, \delta_{kl}.$$

From this expression it is apparent that when $\mathscr{M}_1(\mathscr{A}_2)$ yields $a_k^{(2)}$, a simultaneous $\mathscr{M}_1(\mathscr{A}_1)$ would yield $a_k^{(1)}$. Hence we have an

$\mathscr{M}_2(\mathscr{A}_1)$ scheme: To measure $\mathscr{A}_1$, simply measure $\mathscr{A}_2$; if $a_k^{(2)}$ results, then $a_k^{(1)}$ is regarded as the result of $\mathscr{M}_1(\mathscr{A}_1)$.

Suppose $\mathscr{B}_1$, like $\mathscr{A}_2$, also has an established operational definition. Now, since the $\mathscr{M}_2(\mathscr{A}_1)$ just outlined involves no interaction with S_1, we may perform $\mathscr{M}_1(\mathscr{B}_1)$ simultaneously with $\mathscr{M}_2(\mathscr{A}_1)$ and thereby jointly measure noncommuting observables $\mathscr{A}_1$ and $\mathscr{B}_1$. Once again von Neumann's theorem is contradicted.

6. Strong Correspondence—the Root of Quantum Inconsistencies

Three conclusions may be drawn from the last two sections. (1) The standard quantum postulates (P1S, etc.) rigorously imply that noncommuting observables are incompatible. (2) The same postulates permit empirical arrangements which must be regarded as legitimate schemes for the simultaneous measurement of noncommuting observables, provided the term "measurement" is used in its normal sense. (3) *Hence the standard postulates of quantum theory are inconsistent.* We must therefore reexamine the axiomatic basis of von Neumann's simultaneous measurability theorem in order to discover the false hypothesis which enables the rigorous deduction of this false theorem.

Any theory about the simultaneous measurement of several observables, from axioms referring only to measurements of single observables $\mathscr{M}_1(\mathscr{A})$, requires the notion of compound observable (Secs. 2 and 4); and this concept is subject to consistency conditions (C_1) and (C_2), which would have to be satisfied by any operator corresponding to such a compound observable.

Let us examine the correspondence $\mathscr{A} + \mathscr{B} \leftrightarrow A + B$. Condition (C_1) alone *implies* this rule.[24] Explicitly, P1S guarantees the existence of an operator corresponding to the observable $\mathscr{A} + \mathscr{B}$; that operator would satisfy (C_1) and (C_2). But (C_1) for $\mathscr{A} + \mathscr{B}$ can be satisfied by only *one* operator, namely $A + B$. Condition (C_2), therefore, *need not be used at all.* This observation provides an important clue in our search for the false hypothesis in question.

We ask: Is $\mathscr{A} + \mathscr{B}$ truly an observable? If not, P1S cannot be invoked to assure the existence of an operator counterpart. To answer the question, we recall the last example of Sec. 5. It showed how two systems S_1 and S_2 in a realizable state could be used to construct an appropriate rule of correspondence for simultaneous $\mathscr{M}_1(\mathscr{A}_1)$ and $\mathscr{M}_1(\mathscr{B}_1)$. Now the experimenter is obviously free to add the two results; hence it is apparent that $\mathscr{A}_1 + \mathscr{B}_1$ is observable. *Therefore*, if P1S is true, *there must exist an operator S such that $\mathscr{A} + \mathscr{B} \leftrightarrow S$* in general.

The following simple, contravening example defeats this claim. Let system S_1 be a "spin" whose relevant states and operators span a two-dimensional spinor space. For our noncommuting observables $\mathscr{A}_1$, $\mathscr{B}_1$, we take x- and z-components of spin, $\mathscr{S}_x$ and $\mathscr{S}_z$. Thus, in the Pauli representation,

$$\mathscr{A}_1 = \mathscr{S}_x \leftrightarrow \frac{\hbar}{2}\begin{pmatrix} 0 & 1 \\ 1 & 0 \end{pmatrix}, \qquad \mathscr{B}_1 = \mathscr{S}_z \leftrightarrow \frac{\hbar}{2}\begin{pmatrix} 1 & 0 \\ 0 & 1 \end{pmatrix}.$$

Now simultaneous measurements of $\mathscr{A}_1$ and $\mathscr{B}_1$ employing the correlation with the auxiliary system S_2 will, by Th. 4, always yield one of the eigenvalue pairs: $(\hbar/2, \hbar/2)$, $(\hbar/2, -\hbar/2)$, $(-\hbar/2, \hbar/2)$, $(-\hbar/2, -\hbar/2)$. Hence only the three values $\hbar, 0, -\hbar$ are possible for results of measuring $\mathscr{A}_1 + \mathscr{B}_1$. To use the set notation of Sec. 2,

$$\mathscr{N}(\mathscr{A}_1 + \mathscr{B}_1) = \{-\hbar, 0, \hbar\};$$

and by condition (C_2), if $\mathscr{A}_1 + \mathscr{B}_1 \leftrightarrow S$, quantum mechanics would be self-contradictory unless

$$(C_2) \quad \mathscr{E}(S) \subseteq \mathscr{N}(\mathscr{A}_1 + \mathscr{B}_1).$$

However, (C_1) must also be satisfied by S. As shown in Sec. 4, the only S meeting this requirement is

$$S = \frac{\hbar}{2}\begin{pmatrix} 0 & 1 \\ 1 & 0 \end{pmatrix} + \frac{\hbar}{2}\begin{pmatrix} 1 & 0 \\ 0 & -1 \end{pmatrix} = \frac{\hbar}{2}\begin{pmatrix} 1 & 1 \\ 1 & -1 \end{pmatrix}.$$

But an elementary calculation reveals that the eigenvalues of this operator are $\hbar/2^{1/2}$, $-\hbar/2^{1/2}$; in other words, the set $\mathscr{E}(S) =$

$\{-\hbar/2^{1/2}, \hbar/2^{1/2}\}$. Comparing $\mathcal{N}(\mathcal{A}_1 + \mathcal{B}_1)$ and $\mathcal{E}(S)$, with $\mathcal{A}_1 + \mathcal{B}_1 \leftrightarrow S$, one finds that

$$\mathcal{N}(\mathcal{A}_1 + \mathcal{B}_1) \cap \mathcal{E}(S) = \varnothing.$$

We conclude that the only operator S capable of satisfying (C_1) does not satisfy (C_2).

To summarize: $\mathcal{A}_1 + \mathcal{B}_1$ is demonstrably observable. Axiom P1S then ensures the existence of $S \leftrightarrow \mathcal{A}_1 + \mathcal{B}_1$. If the quantal axioms are consistent, that S must satisfy both (C_1) and (C_2). The *unique* S which satisfies (C_1) violates (C_2). Hence the axioms P1S and P2 are inconsistent.

This conclusion was already suggested at the beginning of this section upon confrontation of von Neumann's theorem with the counterexamples of Sec. 5. But we now see where the difficulty lies among the initial hypotheses leading to that theorem. The theorem is false because P1S—*strong correspondence*—proclaims the existence of operator-observable correspondences which cannot exist in harmony with the remaining postulates. Thus the axiom set—P1S, P2—must be altered. The modification we propose is to replace P1S by P1. Further corroborative evidence for the need of this change (e.g. Temple's theorem) is presented elsewhere.[25]

7. The Consequences of Weak Correspondence

The suggestion that strong correspondence be abandoned is not altogether welcome, primarily because quantum theory would suffer a certain loss of universality as it will not cover all empirical procedures. The historian may console himself in thinking that a fuller range of applicability may be provided by future theoretical discoveries.

At the moment we ask the nonspeculative question: What effect does the replacement of strong by weak correspondence have on the principal quantum theorems? Clearly Th. 1, for example, which implies that every real linear functional $m(A)$ of the Hermitian operators may be expressed in the form Tr (ρA), is quite independent of the physical problem as to whether operators can be found to represent all observables; all that matters is that the operators which *are* involved do represent observables. Within the mathematical framework, operational definitions are irrelevant, and quantum mechanics is a set of mathematical objects subject to given rules.* Their application to the world is made possible by the discovery of rules of correspondence with the *P*-field of experience.[26] None of the parts of linear algebra which form the foundation of quantum theory will be affected by the elimination of strong correspondence. In fact, a careful search through quantum theory for a proposition dependent upon strong correspondence convinced the present writers that no basic theorem involving the analysis of ensembles, statistics of measurement results, etc., requires P1S rather than P1 in its proof.

If only weak correspondence is adopted, the physicist cannot demand in a priori fashion that the mathematician furnish an F for every one of his $\mathscr{F}$'s. Given an observable $\mathscr{F}$, the operator algebra is not *expected* to produce an F; instead, it is simply *asked* whether or not F does exist such that $\mathscr{F} \leftrightarrow F$. In short, what were formerly regarded as "correspondence theorems" are now interpreted as tests of validity for proposed correspondences.

We now offer a summary of the correct interpretation of the theorems of this kind which were mentioned in previous sections:

(1) $\mathscr{A} + \mathscr{B} \leftrightarrow A + B$: (C_1) uniquely determines the operator $A + B$ but (C_2) is often violated. The correspondence is therefore not generally valid.

(2) $\mathscr{A}\mathscr{B} \leftrightarrow \frac{1}{2}(AB + BA)$: Von Neumann's theorem (Sec. 4), in the proof of which this correspondence was central, actually demonstrates that this correspondence can apply only to commuting operators (in which case it takes the simple form $\mathscr{A}\mathscr{B} \leftrightarrow AB = BA$).

(3) What corresponds to $\mathscr{A}\mathscr{B}\mathscr{C}$? Temple's theorem exhibits the ambiguities inherent in this triple product when P1S is assumed.[27]

* Among these are tacit rules concerning the construct $\mathscr{M}_1(\mathscr{A})$ which give meaning to the primitive term observable as it appears in P1.

It is perhaps instructive to consider a simple example which illustrates why consistency condition (C_2) required only $\mathscr{E}(F) \subseteq \mathscr{N}[\mathscr{F}(\mathscr{A}, \mathscr{B})]$ and not the equality of these two sets. Suppose $\mathscr{A} = \mathscr{L}_z^2$, $\mathscr{B} = \hbar\mathscr{L}_z$, where $\mathscr{L}_z$ is the z-component of orbital angular momentum, $\mathscr{L}_z \leftrightarrow L_z = (\hbar/i)\, \partial/\partial\varphi$. $\mathscr{A}$ and $\mathscr{B}$ are measured simultaneously, and the results are added together. The set of all possible results of this procedure, $\mathscr{N}(\mathscr{A} + \mathscr{B})$, is given by $\mathscr{N}(\mathscr{L}_z^2 + \hbar\mathscr{L}_z) = \{m^2\hbar^2 + n\hbar^2\}$, since $\mathscr{E}(L_z) = \{m\hbar\}$. But the eigenvalues of $L_z^2 + \hbar L_z$ comprise the set $\mathscr{E}(L_z^2 + \hbar L_z) = \{k(k+1)\hbar^2\}$, which is only a subset of $\mathscr{N}(\mathscr{A} + \mathscr{B})$. The reason for this inequality is easily understood if postulate (J) of Sec. 3 is recalled. Any measurement of the observables $\mathscr{L}_z^2$ and $\hbar\mathscr{L}_z$ must yield results correlated in the same manner as would be the results of a trivial joint measurement of these observables. One such joint measurement involves simply measuring $\mathscr{L}_z$ and evaluating $\mathscr{L}_z^2 + \hbar\mathscr{L}_z$. But that procedure can yield only numbers in the set $\{k(k+1)\hbar^2\} = \mathscr{E}(L_z^2 + \hbar L_z)$. This demonstration merely affirms the consistency of (J), with the postulated correspondence $\mathscr{F}(\mathscr{C}) \leftrightarrow \mathscr{F}(C)$.

Elementary treatments of quantum mechanics occasionally employ correspondences (1) and (2) as if they represented a universal method of "deriving" quantum operators from classical functions. Since (1) and (2) are false for most $\mathscr{A}$ and $\mathscr{B}$, it is evident that so-called "quantization" schemes based upon (1) and (2) are at best memory aids taking advantage of our familiarity with classical mechanics. (Cf. Ref. 9.)

While replacement of P1S by P1 has no effect whatsoever on the normal applications of the theory to experiment, this revision does have considerable theoretical and philosophical significance. Primarily it shows that von Neumann's simultaneous measurability theorem is a correct mathematical theorem physically misinterpreted as a restriction on measurability. It turns out to be a *reductio ad absurdum* proof that the correspondence $\mathscr{A}\mathscr{B} \leftrightarrow \frac{1}{2}(AB + BA)$ is false unless $[A, B] = 0$; in other words, a proof that $[A, B] = 0$ is a necessary condition for the validity of $\mathscr{A}\mathscr{B} \leftrightarrow \frac{1}{2}(AB + BA)$.

Hence any physical or metaphysical idea motivated by, or founded upon, the concept of incompatibility now requires careful reexamination. Two very common propositions based on incompatibility are the following: (1) Because noncommuting observables are in principle not simultaneously measurable, it is meaningless to contemplate joint probability distributions of quantal measurement results. (2) Since any proposition about the outcome of simultaneous measurements of noncommuting observables is meaningless, a new system of logic is required for quantum physics.

(1) When the incompatibility doctrine has been discarded, there remains no a priori restraint against the study of joint distributions. (For a systematic study of such distributions, cf. Ref. 9.)

(2) At the end of Sec. 4, we indicated how incompatibility led to the notion that quantum mechanics requires a new, "nondistributive" logic, i.e. a system which does not involve the law

$$\mathscr{P} \cap (\mathscr{Q} \cup \mathscr{R}) = (\mathscr{P} \cap \mathscr{Q}) \cup (\mathscr{P} \cap \mathscr{R}),$$

which merely expresses an idea most physicists—including quantum theorists "off duty," to use Landé's phrase—regard as "common sense." The problem was that propositions $\mathscr{P}$, $\mathscr{Q}$, and $\mathscr{R}$ can be given for which there does exist an Hermitian operator corresponding to the left member but there is not one for the right member. Apart from the esoteric context in which it is cast, this problem is not different from the difficulty encountered with the correspondence $\mathscr{A} + \mathscr{B} \leftrightarrow S$. Just as an appropriate S exists only when $[A, B] = 0$, similarly a D exists such that $\mathscr{P} \cap \mathscr{Q} \leftrightarrow D$ only when $[P, Q] = 0$. When $[P, Q] \neq 0$, it simply means that the compound proposition $\mathscr{P} \cap \mathscr{Q}$ has no operator representative D. Naturally it is then impossible to write down an operator counterpart to the distributive law; but this does not make the law wrong!

There are other interesting implications with respect to the "microcausality principle." They are discussed in Ref. 9.

8. A Search for "Simple" Simultaneous Measurements

It is shown elsewhere [28–30] that attempts to approach the study of quantum joint probabilities of noncommuting observables via more or less natural random-variable techniques are thwarted at some stage by ignorance of, or perhaps even the nonexistence of, operators corresponding to compound observables. It is therefore desirable to develop a method for examining simultaneous measurements which does not depend on unknown operator-observable correspondence rules. To do this, we return to the general ideas concerning quantum measurement which were reviewed in Sec. 1. It was seen there that the primitive classical notion of possession ("System S has $\mathscr{A}$-value a_k") is superseded by the primitive quantal measurement construct $\mathscr{M}_1$ ("If $\mathscr{M}_1(\mathscr{A})$ is performed on system S, the value a_k will result with probability . . ."). While a theoretical explanation of measurement processes in classical physics involved relations among possessed attributes, a quantum theory of measurement at best describes connections among the unanalyzable $\mathscr{M}_1$'s. On the other hand, *statements* of such connections and associated empirical procedures constitute the usual scientific concept of measurement, or measurement scheme (operational definition, the epistemic correspondence rule[31]). To signalize the logical distinction, we have designated the latter class of constructs, which form part of the *theoretical* structure of our problem, by $\mathscr{M}_2$. They were exemplified in Sec. 5.

Thus far we have shown by way of examples that there *are* procedures which permit an assignment of values to pairs of noncommuting observables. Our present aim goes beyond such indications; it is to clarify within the context of measurement *theory*, as presented in the foregoing pages, how such empirical operations function as parts of the complete mathematical structures. We shall see that certain kinds of $\mathscr{M}_2$ are free from theoretical difficulties, while others seem to generate internal contradictions.

Because every physical process—hence any measurement scheme, single or joint—has a quantum-theoretical description, it seems reasonable that, whatever the correct joint probabilities are, they should be *derivable* within the framework of a quantum theory of $\mathscr{M}_2$. That is, if a given procedure $\mathscr{M}_2(\mathscr{X}, \mathscr{P})$ is to be regarded as a method for simultaneous measurement of $\mathscr{X}$ and $\mathscr{P}$, the scheme must be certified by a theory establishing relations between $\mathscr{M}_1(\mathscr{X})$, $\mathscr{M}_1(\mathscr{P})$ and whatever "direct meter readings" are used as the basis for inference of simultaneous $\mathscr{M}_1(\mathscr{X})$- and $\mathscr{M}_1(\mathscr{P})$-results; from this analysis it should be possible in principle to find the probability for the occurrence of those "meter readings" that imply any given pair of $\mathscr{X}$- and $\mathscr{P}$-values. This measurement-theoretical approach to the joint-probability problem bypasses the difficulty associated with the operator-observable correspondence, which obstructed the methods reviewed earlier. All this will be clarified below by explicit examples.

To develop these ideas further, we next distinguish two kinds of $\mathscr{M}_2$-concepts: (1) simple or type A and (2) historical or type B. This distinction will later turn out to have considerable bearing on the problem of compatibility.

(1) A simple $\mathscr{M}_2$ begins with system S in an arbitrary state ρ_{t_0} at some specified time t_0 and demonstrates how some single operation upon S eventually leads to numbers from which may be inferred $\mathscr{M}_1$-results to be associated with S in state ρ_{t_0}. It is to be especially noted that the state of S *before* t_0 is completely irrelevant. We shall also refer to this class of measurement as belonging to type A.

(2) An *historical* $\mathscr{M}_2$-theory also seeks to certify some operation as a bona fide supplier of numbers which can be meaningfully interpreted as $\mathscr{M}_1$-results for S in state ρ_{t_0}. However, unlike the simple type A, the historical $\mathscr{M}_2$-theory cannot be worked out without detailed information concerning the structure of ρ_{t_0}. Such information might be deduced from facts about the history of the system, e.g. its state at some earlier time $t_1 < t_0$ plus its physical environment between t_1 and t_0. An example of each type appeared in Sec. 5: the simple time-of-flight $\mathscr{M}_2(\mathscr{P})$ of type A and the historical time-of-flight $\mathscr{M}_2(\mathscr{X}, \mathscr{P})$ of type B.

Physically, the $\mathscr{M}_2$-theories of type A have been of greatest interest, because they represent the idea of measurement in its most primitive form, as a process applicable to a system at any instant independently of its past. Similarly, in quantum mechanics the language of $\mathscr{M}_1$'s tends to presuppose that measurements are performed upon systems in states which are simply given without details as to the actual method of preparation. Accordingly, $\mathscr{M}_2$-schemes for single observables (or commuting sets of observables) have been of type A. One might therefore be tempted to seek a simple $\mathscr{M}_2$-theory covering the simultaneous measurement of several noncommuting observables. However, in view of the fact that both examples of simultaneous measurement given in Sec. 5—the time-of-flight $\mathscr{M}_2(\mathscr{X}, \mathscr{P})$ and the use of two systems already correlated at the time of interest—were of type B, there is so far no reason to expect any *simple* theory for simultaneous measurement.

Elsewhere[32] we have examined two fairly general procedures which, at the outset, seem to be altogether plausible methods for achieving simultaneous type A measurement of two noncommuting observables. In both cases, theoretical obstacles eventually arose, and this may be interpreted as evidence that quantum theory does perhaps forbid *type A* simultaneous measurements. Deeper reasons to anticipate such a theoretical prohibition have also been explored.[33]

These examples furnish partial evidence for this proposition:

(0)
Simultaneous type A measurements of noncommuting observables are theoretically impossible.

Of course, merely citing two unsuccessful attempts to develop a simple $\mathscr{M}_2(\mathscr{A}, \mathscr{B})$ does not prove (0); nevertheless there appears, for the first time in the present study, good reason to suspect that quantum theory may indeed place some restriction upon joint measurability. If so, the qualification will not be a sweeping mandate to the effect that $\mathscr{M}_2(\mathscr{A}, \mathscr{B})$ is generally impossible, since that common version was refuted in Sec. 5 by counterexamples. Rather,

(0) would mean only this: Given at time t_0 a system S *of unknown history*, it is impossible to devise an operation $\mathscr{M}_2(\mathscr{A}, \mathscr{B})$ which leads to numbers (a_k, b_l) interpretable as $\mathscr{M}_1(\mathscr{A})$- and $\mathscr{M}_1(\mathscr{B})$-results for time t_0.

References

1.
A. Landé, *New Foundations of Quantum Mechanics* (Cambridge University Press, 1965), p. 124.
2.
J. L. Park, Am. J. Phys. **36**, 211 (1968).
3.
J. L. Park, Phil. Sci. **35**, pt. I, 205; pt. II, 389 (1968).
4.
H. Margenau, Phil. Sci. **30**, 6 (1963).
5.
H. Margenau, Phys. Rev. **49**, 240 (1936).
6.
J. von Neumann, *Mathematical Foundations of Quantum Mechanics*, trans. by R. T. Beyer (Princeton University Press, 1955).
7.
Ref. 4.
8.
Ref. 3.
9.
J. L. Park and H. Margenau, Intern. J. Theoret. Phys. **1**, 211 (1968). This paper contains a fuller account and more detailed proofs of many of the results set forth in the present article.
10.
H. Margenau, Phys. Today **7**, 6 (1954).
11.
H. Margenau, Phil. Sci. **4**, 352–356 (1937).
12.
G. C. Wick, A. S. Wightman, and E. P. Wigner, Phys. Rev. **88**, 101 (1952).
13.
Ref. 6, pp. 313–316. Cf. also A. M. Gleason, J. Math. Mech. **6**, 885 (1957).
14.
See, for instance, H. J. Groenewold, Physica **12**, 405 (1946); J. R. Shewell, Am. J. Phys. **27**, 16 (1959).
15.
Ref. 9.
16.
J. von Neumann, Ann. Math. **32**, 191 (1931).

17.
Ref. 9.
18.
Ref. 6, pp. 225–230.
19.
E.g.: G. W. Mackey, *Mathematical Foundations of Quantum Mechanics* (W. A. Benjamin, 1963), p. 70; I. E. Segal, Ann. Math. **48**, 930–948 (1947); V. S. Varadarajan, Commun. Pure Appl. Math. **15**, 189–217 (1962).
20.
E.g.: G. Birkhoff and J. von Neumann, Ann. Math. **37**, 823 (1936); H. Reichenbach, *Philosophic Foundations of Quantum Mechanics* (University of California Press, 1944); C. Piron, Helv. Phys. Acta **37**, 439 (1964).
21.
R. P. Feynman, *Quantum Mechanics and Path Integrals* (McGraw-Hill, 1965), pp. 96–98.
22.
W. Heisenberg, *The Physical Principles of the Quantum Theory* (University of Chicago Press, 1930), p. 20.
23.
Ibid.
24.
Ref. 9.
25.
Ibid.
26.
Cf. H. Margenau, *The Nature of Physical Reality* (McGraw-Hill, 1950), pp. 171–177.
27.
Ref. 9.
28.
H. Margenau and R. N. Hill, Progr. Theoret. Phys. (Kyoto) **26**, 727 (1961).
29.
H. Margenau, Ann. Phys. (N.Y.) **23**, 469 (1963).
30.
L. Cohen, J. Math. Phys. **7**, 781 (1966); H. Margenau and L. Cohen in *Quantum Theory and Reality*, ed. by M. Bunge (Springer-Verlag, 1968).
31.
H. Margenau, Phil. Sci. **2**, 1 (1935).
32.
Ref. 9.
33.
Ibid.

Six **Alfred Kastler**

How the Landé Factor of an Atom Can Be Changed by Putting the Atom in a Radiofrequency Bath

1. Review of General Definitions

The *g-factor* is a characteristic property of a paramagnetic energy state of an atom. It determines the amount of splitting into Zeeman sublevels by an external magnetic field H_0 applied to the atom.

For singlet states this splitting between adjacent m-sublevels is the normal Zeeman splitting, given by

$$\Delta E = \frac{e\hbar}{2m_0} H_0, \qquad [6.1]$$

where $\mu_B = e\hbar/(2m_0)$ is the Bohr magneton.

For other states we have to write

$$\Delta E = g \frac{e\hbar}{2m_0} H_0, \qquad [6.2]$$

where g is the factor introduced by Landé.

The *gyromagnetic ratio*, which is the ratio of the magnetic moment to the angular moment of the atom, is defined as

$$\gamma = g \frac{e}{2m_0}, \qquad [6.3]$$

and the *Larmor precession frequency* around the field H_0 is given by

$$\omega_0 = \frac{\Delta E}{\hbar} = \gamma H_0 . \qquad [6.4]$$

2. Measurement of the Landé *g*-Factor by the Hanle Method

In 1924, W. Hanle[1] developed an elegant optical method for measuring the Larmor precession frequency ω_0 of an atomic state. By irradiation of a monatomic vapor at low vapor pressure with optical resonance radiation, a bulk electric or magnetic dipole moment in a given space direction can be produced in the vapor. As a result of this induced anisotropy, the fluorescence light reemitted by the vapor appears polarized. Applying a small magnetic field H_0 in a direction perpendicular to this induced moment, the effect on this moment of the Larmor precession around this field H_0 can be observed. Let $0x$ be the direction of the dipole moment induced by the light and $0z$ the direction of field H_0. The moment produced at constant rate in time by the light is spread by the Larmor precession to form a fan in the plane perpendicular to H_0. At the same time, this precessing moment decays with a characteristic decay time τ. In steady-state conditions the figure formed by the fan in the plane $x0y$ depends on the value of the product $\omega_0\tau$.

Figure 6.1 shows the shape of the fan for three cases:

(a) $\omega_0\tau \ll 1$, (b) $\omega_0\tau = 1$, (c) $\omega_0\tau \gg 1$.

A simple calculation shows that the component M_x of the moment in the $0x$ direction is given by

$$M_x = \frac{M_0}{1 + (\omega_0\tau)^2}, \qquad [6.5]$$

where M_0 is the steady-state moment attained at $H_0 = 0$.

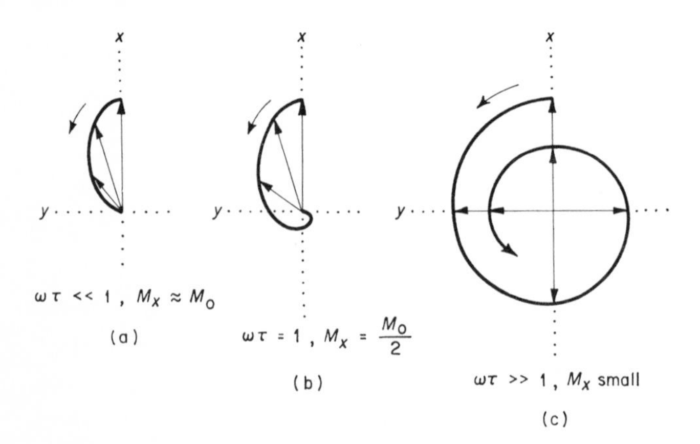

Figure 6.1 Hanle experiment: Larmor precession of the induced transverse moment M_x.

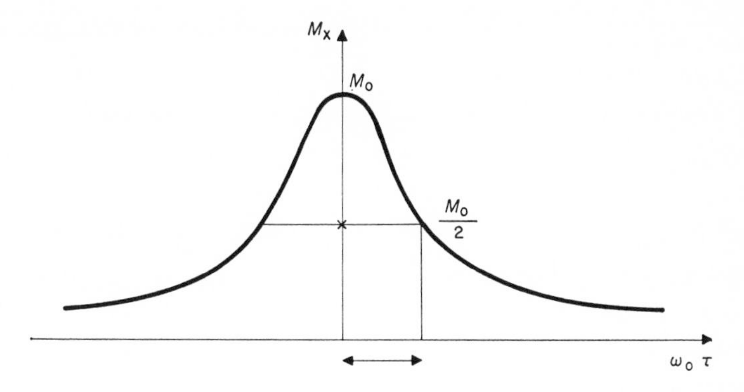

Figure 6.2 Hanle curve.

Figure 6.2 shows the theoretical *Hanle curve*, which is Lorentz-shaped. The *half-width* of this curve (abscissa for which the ordinate has half its maximum value) corresponds to $\omega_0 \tau = 1$. According to Eq. [6.4] the associated field value is

$$H_0 = \frac{2m_0}{e\tau} \frac{1}{g}.$$

For a given decay time τ the half-width of the Hanle curve is inversely proportional to the Landé factor g.

In the original experiment performed by Hanle, the dipole moment induced by irradiation with linearly polarized light was an electric dipole moment produced by excitation of the 6^3P_1 state of the mercury atom (for even isotopes of mercury). In this case, the decay time τ in formula [6.5] is identical with the lifetime of the excited atomic state.

Hanle's method has been combined with the optical pumping technique to induce, by irradiation with circularly polarized resonance light, a magnetic dipole moment in the ground states of atoms. For this case τ is the relaxation time of the ground state.

In this manner, Hanle curves have been traced for ground states of atoms having diamagnetic electron configurations ($^1S_0$ states) but exhibiting nuclear paramagnetism. This approach has been applied especially to Cd^{113} and to Hg^{199}.[2] In both examples the nuclear spin is $I = \frac{1}{2}$ and the ground state has only one Zeeman interval between the states $m = -\frac{1}{2}$ and $m = +\frac{1}{2}$. The Hanle curve for Hg^{199} is shown in Fig. 6.4(a); it is the curve corresponding to $V_1 = 0$.

3. The Landé Factor of Atoms in a Radiofrequency Bath

In the preceding section we have described how the Hanle curve is obtained for free atoms—the atoms of a very dilute monatomic vapor. These atoms are irradiated with a light which induces a dipole moment M_x in the $0x$ direction, and the change of this moment by a steady small magnetic field H_0 applied in the $0z$ direction is observed. See Fig. 6.3.

This experiment has been modified by C. Cohen-Tannoudji and Haroche[3] in the following way. By a radiofrequency coil surrounding a cell filled with vapor of isotope Hg^{199}, a rf field $H_1 \cos \omega t$ is steadily applied in the $0x$ direction to the atoms of the vapor. In the absence of any field H_0, the moment induced by the light, M_x, is not affected by the rf field. (Note that the rf field is parallel to M_x.) In this case a Larmor precession around H_1 will not change the moment M_x. If the rf field, instead of being parallel, is per-

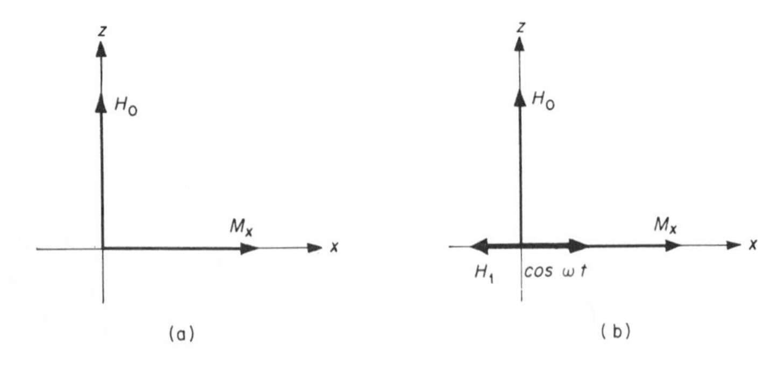

Figure 6.3 Change of dipole moment M_x by magnetic fields.
(a) With a steady field H_0 in $0z$ direction.
(b) With a steady field H_0 in $0z$ direction and a radiofrequency field in $0x$ direction.

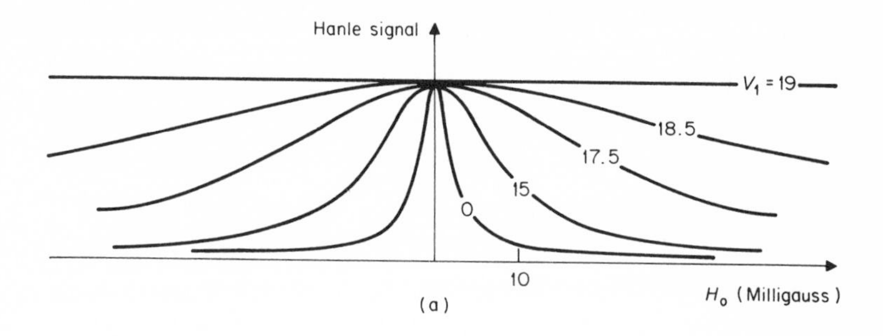

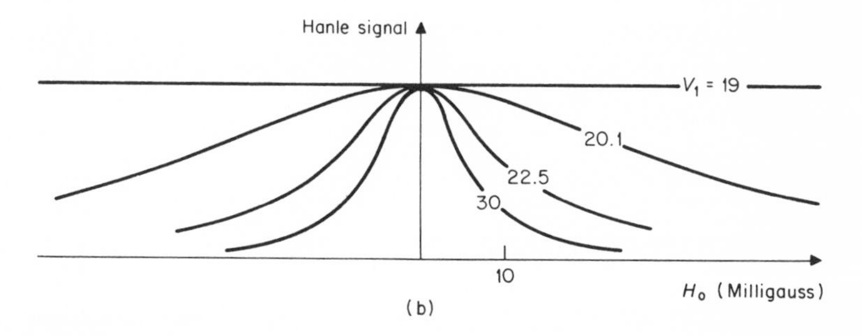

Figure 6.4 Hanle curves for increasing values of the voltage V_1 applied to the rf circuit.
(a) $0 \leq V_1 \leq 19$.
(b) $19 \leq V_1 \leq 30$.

pendicular to M_x, a change will occur. We have then to deal with the Fermi-Rasetti experiment.[4]

If now the steady field H_0 is applied, the Hanle curve appears drastically changed. Figure 6.4 shows Hanle curves for increasing values of the amplitude H_1 of the rf field. For each of these curves, V_1 indicates the voltage applied to the rf circuit; its value is proportional to H_1. The V_1 scale corresponds to the H_1 scale in arbitrary units. We see that for increasing values of H_1 (up to $V_1 = 19$), the Hanle curve becomes larger and larger. This means that the g-factor of the atom becomes smaller and smaller. For $V_1 = 19$, the Hanle curve is infinitely large, and at still higher values of H_1 a narrowing of the Hanle curve is observed.

In all these experiments, the following conditions are fulfilled:

$\omega \gg \omega_0 = \gamma H_0$ and $H_1 \gg H_0$ or $\omega_1 = \gamma H_1 \gg \omega_0$.

Figure 6.5 shows the summary of the results: The abscissa corresponds to $\omega_1/\omega = \gamma_0 H_1/\omega$, where ω is the circular frequency and H_1 the amplitude of the rf field and $\gamma_0 = g_0 e/(2m)$ is the normal gyromagnetic ratio of the atom. The ordinate shows the ratio g_i/g_0, g_0 being the normal g-factor of the free atom and g_i the modified g-factor of the atom surrounded by the rf bath.

At point A the g-factor has become zero, so that the application of field H_0 produces no Zeeman splitting. Between points A and B, the sign of the g-factor is reversed. The experimental points lie on a curve represented by the formula

$$\frac{g_i}{g_0} = J_0\left(\frac{\omega_1}{\omega}\right),$$

where J_0 is the Bessel function of zeroth order.

The theory of this effect has been developed by Cohen-Tannoudji and Haroche[5] and is based on the idea that the atom with the surrounding photon-field has to be treated as a single quantized system.

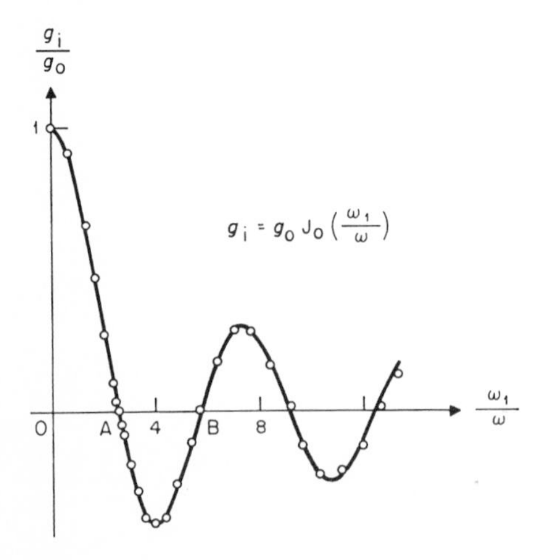

Figure 6.5 Summary of results.

If one studies the energy levels of this whole system as a function of H_0, it is found that for certain values of ω_1 the shape of the levels near $H_0 = 0$ is zero. For these values of ω_1, which correspond to the zeros of the Bessel function $J_0(\omega_1/\omega)$, the interaction between the atoms and the rf field has canceled the magnetic moment of the whole system. This interaction can be described in terms of *virtual* absorptions (or induced emissions) of rf quanta by the atoms, which do not conserve energy. In agreement with the uncertainty relation, photons can be absorbed by an atom if they are reemitted after a very short time interval.

There is an intimate relation between these absorption and reemission processes, which result in a substantial change of the *g*-factor of the atom, and the emission and reabsorption processes of virtual photons that lead to the anomalous *g*-factor of the electron.[6]

References

1.
W. Hanle, Z. Physik **30**, 93 (1924); Erbeg. Exakt. Naturw. **4**, 214 (1925). See also A. C. G. Mitchell and M. W. Zemansky, *Resonance Radiation and Excited Atoms* (Cambridge University Press, 1934), chap. 5.
2.
J. C. Lehmann and C. Cohen-Tannoudji, Compt. Rend. **258**, 4463 (1964).
3.
C. Cohen-Tannoudji and S. Haroche, Compt. Rend. **262**, 268 (1966).
4.
E. Fermi and F. Rasetti, Z. Physik **33**, 246 (1925).
5.
Ref. 3.
6.
F. Bloch, Physica **19**, 821 (1953); J. Schwinger, Phys. Rev. **73**, 416 (1948); P. Kusch and H. M. Foley, Phys. Rev. **74**, 250 (1948); S. H. Koenig, A. G. Prodell, and P. Kusch, Phys. Rev. **88**, 191 (1952); R. P. Feynman, *Quantum Electrodynamics* (W. A. Benjamin, 1962), lectures 26 and 27; S. Haroche, C. Cohen-Tannoudji, C. Andoin, and J. P. Schermann, Phys. Rev. Letters **24**, 816 (1970); S. Haroche and C. Cohen-Tannoudji, Phys. Rev. Letters **24**, 974 (1970).

Seven **David Bohm**

Space-Time Geometry as an Abstraction from "Spinor" Ordering

1. Introduction

Ever since the work of Pauli on "spin" and its relativistic generalization by Dirac, it has been implicit in quantum theory that the notion of "spinor" is in some sense more basic than the ordinary geometrical notion of "vector" (and "tensors" that can be built out of vectors). For it is possible to express any vector uniquely as bilinear combinations of spinors, while no such a unique inverse correspondence is possible which would express spinors in terms of vectors.

Of course, there is a well-known way of obtaining a double-valued correspondence between spinor and vector quantities. To obtain the correspondence, we let $F_{\mu\nu}$ represent an antisymmetric tensor and $\tilde{F}_{\mu\nu}$ be its dual, such that $F_{\mu\nu}\tilde{F}^{\mu\nu} = 0$ and $F_{\mu\nu}F_{\lambda\alpha}\varepsilon^{\mu\nu\alpha\lambda} = 0$ (where $\varepsilon^{\mu\nu\alpha\lambda}$ is the antisymmetric tensor, all of whose indices are 1, -1, or 0). Then, if V^{μ} is a null four-vector, satisfying $V^{\mu}F_{\mu\nu} = 0$ and $V^{\mu}\tilde{F}_{\mu\nu} = 0$, one can write

$$V_{\mu} = \psi^{*}_{A}\sigma^{AB}_{\mu}\psi_{B}$$

$$F_{\mu\nu} = \psi^{A}\varepsilon_{AB}\sigma^{BC}_{\mu\nu}\psi_{C},$$

where ψ^{A} represents a two-component spinor, ε_{AB} is the antisymmetric symbol which is 1, -1, or 0, σ^{AB}_{μ} is a Pauli spin matrix, and $\sigma^{BC}_{\mu\nu}$ can be built out of Pauli spin matrices,

$$(\sigma_{\mu\nu})^{BC} = \tfrac{1}{2}(\sigma_{\mu}\sigma_{\nu} - \sigma_{\nu}\sigma_{\mu})^{BC}.$$

This evidently yields a two-to-one correspondence, in which each vector V_{μ}, and tensor $F_{\mu\nu}$, define a spinor ψ^{A}, *but only up to a sign* (i.e. reversing the sign of ψ^{A} leads to the same V_{μ} and $F_{\mu\nu}$).

One can picture the preceding correspondence by saying that each spinor defines a null four-vector, and a so-called "flag" (i.e. a two-dimensional "plane" element that contains the null four-vector as a line within the plane in question).

It may not seem very important, at first sight, that the correspondence between spinor and vector is not one-valued. Nevertheless (as has been shown elsewhere[1]), it is of crucial significance when one comes to inquire into the *continuity* of structures whose "elements" would be of the kind described above, i.e. "null vectors with flags." Indeed, to make up a continuous structure of such "elements" it has been demonstrated that it is necessary for these "elements" to be parameterized in a one-to-one correspondence with the *spinor*. Thus, vectors are inadequate for the treatment of the "global" topological order of such structures constituted out of "vectors with flags" (though it may perhaps be adequate for the "local order" of these structures).

To understand the very broad issues that are involved in this question, it will be useful to consider the notion that our "geometrical" concepts of space and time are defined through certain kinds of *order* that are in essence implied in the properties attributed to a "quasi-rigid" object. Now, it is known through common experience that this kind of object is, in some sense, "invariant" in its intrinsic properties, under the group of displacements and rotations in space. If we now generalize to space-time, such an object has, of course, to be considered in its pattern of movement and development. (For example, a simple object might be represented by some sort of "world tube" in a Minkowski diagram, as shown in Fig. 7.1).

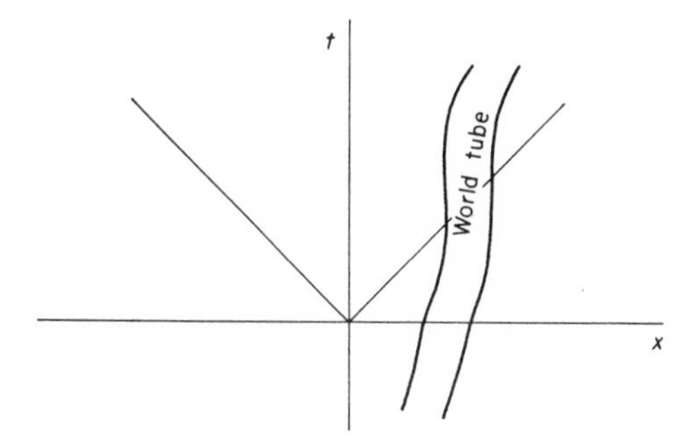

Figure 7.1 World tube in a Minkowski diagram.

The essential content of the theory of relativity is then that these space-time patterns are "invariant" in their intrinsic properties under the Poincaré group (space-time displacement and rotation). This kind of invariance group (whether in space, or in space-time) then implicitly defines a set of *ordered correspondences*. Thus, an "object" or a "pattern of movement" can be displaced in any direction, and the order of results of such a displacement is in a one-to-one correspondence with the orders of the parameters of the relevant displacement group. Similarly, orientations (in space-time) can be ordered in correspondence with the parameters of the Lorentz group.

From the preceding, it can be seen that a considerable part of the content of our "geometrical" concepts lies in the *order* that is defined by the basic terms of description (in this case, the quasi-rigid object or "pattern of movement") and the groups under which the basic terms are, in some sense, invariant.

It will be the aim of this paper to suggest a new notion of order in which the "spinor" plays a fundamental role analogous to that defined for the vector in ordinary geometry. The "spinor" order will be visualized in terms of a set of projectively related hyperplanes. The invariance group of the "spinors" will be seen, however, to be *more general* than the Lorentz group (which latter will be only a subgroup of the full invariance group). Thus, the "spinor" order is in essence not "geometric" in the sense defined before. It is not expressible in terms which take as basic the notion of space-time as a potential locus of "quasi-rigid" structures, which are invariant under the Poincaré group.

This means, for example, that the light cone is no longer regarded as a fundamental element in determining order. Rather, it becomes a formal abstraction (as does indeed any structure constituted out of vectors) from "spinor" order, which is taken as primitive. Such a notion opens the way for new forms of description, in which relativity as a means of expressing space-time order will have only limited relevance (as relativity likewise implied the limited relevance of Newtonian notions of space and time order). Entirely novel possibilities for order are thus suggested, but the detailed discussion of these will be deferred for a later paper.

2. Spinor Ordering in Three Dimensions

Let us begin with a light wave, whose source is taken to be at the origin of coordinates. At time t, the wave front is a sphere, of radius $r = ct$. We now consider what we shall call a "double stereographic projection" of the surface of this sphere onto two planes tangent to the sphere and contacting it at opposite ends of a diameter. We choose these planes parallel to $z = 0$, as shown in Fig. 7.2. Stereographic projections of the point P (with coordinates, x, y, z) are made from opposite ends of the diameter, AB. If $2r\xi_1, 2r\eta_1$ are the coordinates of the projection Q, while $2r\xi_2, 2r\eta_2$ are those of the projection R, then we have

$$\xi_1 = \frac{x}{r - z}, \qquad \eta_1 = \frac{y}{r - z},$$

$$\xi_2 = \frac{x}{r + z}, \qquad \eta_2 = \frac{y}{r + z},$$

where

$$x^2 + y^2 + z^2 = r^2.$$

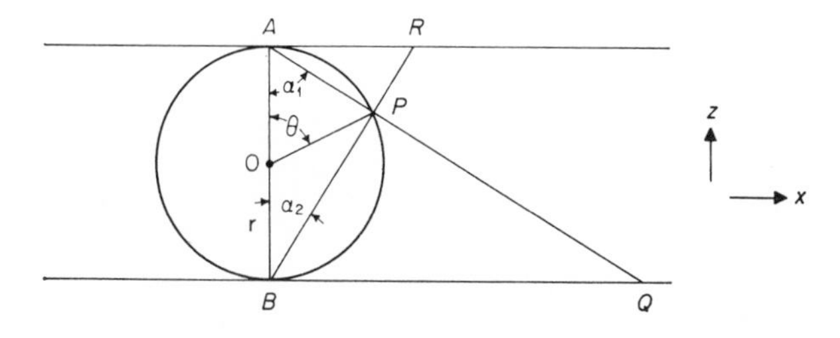

Figure 7.2 Double stereographic projection of a spherical wave front.

To simplify the discussion, we begin with a case for which $y = 0$. We introduce homogeneous (projective) coordinates for ξ_1 and ξ_2:

$$\xi_1 = \frac{\psi_A}{\psi_B}, \qquad \xi_2 = -\frac{\varphi_A}{\varphi_B}.$$

We now consider

$$\xi_1 = \frac{x}{r - z} = \frac{\psi_A}{\psi_B}.$$

If we write

$$x = \psi_A \psi_B \qquad r - z = \psi_B^2$$

$$z = \psi_A^2 \qquad\qquad r = \psi_A^2 + \psi_B^2,$$

it is clear that the vector (x, z) is related to ψ_A and ψ_B exactly as it is related to a spinor, for the special case in which y vanishes (so that the spinor can be taken to be real). Therefore, the "spinor components" ψ_A and ψ_B are projective coordinates of the point (ξ_1, η_1) or, alternatively, of the line AQ, which is in a one-to-one correspondence with this point. To obtain this, we derive the equation of the line AQ by cross-multiplying terms in the equation

$$\frac{x}{r - z} = \frac{\psi_A}{\psi_B}$$

to yield

$$\psi_B x - \psi_A(r - z) = 0.$$

Similarly, we can obtain the equation for the line BR by cross-multiplying

$$\frac{x}{r + z} = -\frac{\varphi_A}{\varphi_B}$$

or

$$\varphi_B x + \varphi_A(r + z) = 0.$$

These two lines intersect at the point P. In order that this point be on a circle of radius r, we must have

$$x^2 = r^2 - z^2 \qquad \text{or} \qquad \frac{x}{r - z} = \frac{r + z}{x}.$$

This implies that

$$\frac{\psi_A}{\psi_B} = -\frac{\varphi_B}{\varphi_A} \qquad \text{or} \qquad \varphi_A\psi_A + \varphi_B\psi_B = 0.$$

In purely formal terms this means that the spinors (ψ_A, ψ_B) and (φ_A, φ_B) correspond to what are called "orthonormal vectors" in a Hilbert space of two dimensions. But in projective geometry it means that the lines AQ and BR are in a certain kind of correspondence which is a special case of what can be called "harmonic." The essential significance of this kind of "harmonic" correspondence is that the two lines are perpendicular to each other, and their intersection defines a *circle*, whose diameter is given by the projection points A and B.

Note that if θ is the angle of the radius OP relative to the z-axis, then AP subtends an angle $\alpha_2 = \theta/2$, while BP subtends $\alpha_1 = (\pi/2) - (\theta/2)$. (In other words, as is well known, the chord subtends only half the angle determined by the radius.) Here, we see the basic origin of the two-to-one correspondence between spinors and geometry: If we consider the general locus of the point P to define a geometrical entity (i.e. a circle), then when the radius vector OP rotates through an angle β, the chords AP and BP rotate respectively through $\beta/2$ and $-\beta/2$. Consequently, if the radius vector rotates through 360° (and thus returns to its original direction), the chords AP and BP rotate only through 180° and are therefore reverse in direction. Thus, there is a two-to-one correspondence between chords and radius vector. This is expressed by the spinor transformation involving the angle $\beta/2$:

$$\psi_A = \psi'_A \cos\frac{\beta}{2} + \psi'_B \sin\frac{\beta}{2},$$

$$\psi_B = \psi'_B \cos\frac{\beta}{2} - \psi'_A \sin\frac{\beta}{2},$$

$$\varphi_A = \varphi'_A \cos\frac{\beta}{2} + \varphi'_B \sin\frac{\beta}{2},$$

$$\varphi_B = \varphi'_B \cos\frac{\beta}{2} + \varphi'_A \sin\frac{\beta}{2}.$$

3. "Spinor" Ordering and Space-Time

In addition to describing rotations in space, such projective correspondences of lines can also describe Lorentz transformations in space-time.

To show how this comes about, we first introduce the van der Waerden "tensorlike" notation for the spinor indices A and B. From the way in which ψ_A and φ_A are defined, it follows that one may write

$$\varphi^A = \varepsilon^{AB}\psi_B.$$

The invariant is then

$$I = \varphi^A\psi_A = \varphi_A\varepsilon^{AB}\psi_B = \varphi_A\psi_B - \varphi_B\psi_A = \text{Det}\begin{pmatrix}\varphi_A & \psi_A \\ \varphi_B & \psi_B\end{pmatrix}.$$

Thus, a transformation that keeps I invariant is one that leaves the determinant of a matrix, such as

$$M_{AB} = \begin{pmatrix}\varphi_A & \psi_A \\ \varphi_B & \psi_B\end{pmatrix},$$

invariant.

In accordance with familiar notation, one can express the coordinates, through a symmetric "spinor" matrix Z_{AB} as

$$Z_{11} = ct - z \qquad Z_{12} = x$$

$$Z_{21} = x \qquad Z_{22} = ct + z.$$

Then

$$\text{Det } Z_{AB} = c^2t^2 - z^2 - x^2 = s^2$$

is the usual relativistic expression for the interval. So all unideterminantal transformations leave s^2 invariant. Therefore, such transformations as "spinor" quantities lead to Lorentz transformations or the ordering of space-time quantities.

These transformations evidently correspond to a *subgroup* of the whole Lorentz group, with operations (L_z, L_x) corresponding to Lorentz transformations along x- and z-axes, and R_{xz} corresponding to rotations in the x, z plane. To include the whole Lorentz group, we need to consider *complex* spinors, which will be discussed in a later paper.

Note that to relate the Lorentz transformation to our earlier treatment of "spinor" order in three dimensions, it is necessary to write $ct = r$, where r is the radius of a spherical wave front, at the time t. Thus, in terms of our original three-dimensional visualization, a Lorentz transformation is to be described as a correspondence of spheres into spheres. *It is not the usual kind of point-to-point correspondence of ordinary geometrical transformations.* Such a correspondence of spheres into spheres was first suggested in the "higher spherical geometry" of Lie and Klein. This notion is indeed very relevant to "spinor" ordering and will be developed in more detail in a later paper.

One can write two basic Lorentz transformations in terms of the parameters z, γ, t. These are

$$z - ct = (z' - ct')e^{\gamma}$$

$$z + ct = (z' + ct')e^{-\gamma},$$

which correspond to a Lorentz transformation in the z, t plane. The "spinor" transformation that brings this about is then

$$\psi_A = \psi'_A e^{\gamma/2}, \qquad \psi_B = \psi'_B e^{-\gamma/2}.$$

A Lorentz transformation in the x, t plane is brought about by the "spinor" transformation

$$\psi_A = \psi'_A \cosh \frac{\gamma}{2} + \psi'_B \sinh \frac{\gamma}{2}$$

$$\psi_B = \psi'_B \cosh \frac{\gamma}{2} + \psi'_A \sinh \frac{\gamma}{2}.$$

Such transformations can be visualized as a more general projective correspondence of lines than that which produces a simple rotation. Thus, for example, consider the lines

$$\psi_B x - \psi_A(ct - z) = 0$$

and

$$\varphi_B x + \varphi_A(ct + z) = 0.$$

Under a Lorentz transformation in the z, t plane, these go over into

$$\psi'_B x - \psi'_A(ct - z) = 0,$$

$$\varphi'_B x + \varphi'_A(ct + z) = 0,$$

$$e^{-\gamma/2}\psi_B x - e^{\gamma/2}\psi_A(ct - z) = 0;$$

$$e^{\gamma/2}\varphi_B x + e^{-\gamma/2}\varphi_A(ct + z) = 0.$$

This is a projective correspondence, in which the points of stereographic projection go from one radius to another so that the point remains on the new sphere that corresponds to the old point (as shown in Fig. 7.3).

4. "Spinor" Order as a Primitive Notion

Now, the fundamentally new step in this work will be to change the notion of what is basic in our *description*. Instead of taking the circle, with its center O and its radius vector OP as basic, we shall

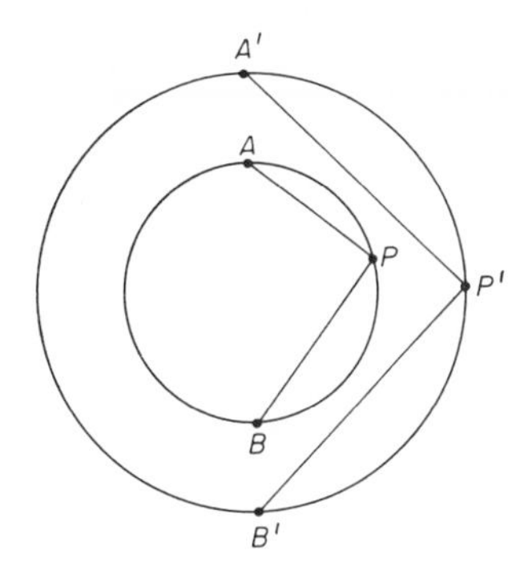

Figure 7.3 Transfer of the points of stereographic projection as a spherical wave front expands from one radius to another.

say that sets of lines through points such as A or B are basic. *The circle whose diameter is AB is then generated by the intersection of a pair of "harmonically" related lines*, through A and B. In other words, "points" (and their loci which delineate geometrical forms and structures) are now *abstracted* as the intersections of certain sets of lines. The *order of the lines is basic*, while the order of the points follows from that of the lines. This is a reversal of the usual relationship of points and lines in geometry.

It follows from this view that the light cone (which is defined in essence by the circular locus of intersection points P) is no longer a basic concept in our theory. Rather, it is a formal abstraction from certain coordinated spinor orders (i.e. of the two intersecting lines AP and BP). Our point of view is that these coordinated spinor orders are to be taken as primitive and that the light cone is to be taken as another order, which is a result of these primitive orders.

One can see an example of how the new spinor order can become relevant by considering a new kind of transformation in which only one of the spinors, ψ_B, is transformed, while the other, ψ_A, is not. Thus, if we write

$$M_{AB} = \xi_A \psi_B,$$

and make a linear transformation, $\psi_B = T^C_B \psi'_C$, we obtain

$$M_{AB} = \xi_A T^C_B \psi'_C = M'_{AC} T^C_B.$$

Now, this transformation is not symmetry-preserving for matrices such as Z_{AB}. To take this into account, let us generalize Z_{AB} by writing

$$Z_{11} = ct - z \qquad Z_{12} = x - s$$

$$Z_{21} = x + s \qquad Z_{22} = ct + z,$$

so that

$$\text{Det } Z_{AB} = c^2t^2 - z^2 - x^2 + s^2.$$

If we restrict ourselves to unideterminantal transformation, we find

$$\text{Det } Z_{AB} = \text{invariant},$$

$$c^2t^2 - z^2 - x^2 + s^2 = \text{invariant}.$$

This transformation is evidently not a Lorentz transformation but something more general. To see what it can mean, let us set the preceding invariant equal to zero. This gives

$$z^2 + x^2 = c^2t^2 + s^2;$$

$s = 0$ corresponds to the light cone. But a fixed value of s that is not equal to zero corresponds to a space-time hyperboloid of revolution. So we have a transformation which is a *correspondence of hyperbolas into each other*. This is evidently a generalization of the spherical correspondence of Lie and Klein. Let us call this generalization "the higher hyperbolic geometry." Such a hyperboloid can be expressed projectively by these ratios:

$$\frac{x - s}{ct - z} = \frac{ct + z}{x + s} = -\frac{\psi_A}{\psi_B}$$

$$\frac{x + s}{ct - z} = \frac{ct + z}{x - s} = -\frac{\xi_A}{\xi_B}.$$

The first of these ratios determines a pair of hyperplanes

$$\psi_B(x - s) = -\psi_A(ct - z)$$

$$\psi_B(ct + z) = -\psi_A(x + s),$$

and these intersect in a line, which we call A. This line is evidently in the hyperboloid $x^2 - s^2 = c^2t^2 - z^2$ (as shown by cross-multiplying these ratios).

The second of these ratios determines another pair of hyperplanes

$$\xi_B(x + s) = -\xi_A(ct - z)$$
$$\xi_B(ct + z) = -\xi_A(x - s),$$

and these intersect in a line, which we call B (which, of course, is also in the hyperboloid).

In fact, A and B are *conjugate lines*, which make up the hyperboloid as a ruled surface. A given point P is defined as the intersection of a pair of conjugate lines A and B.

Transforming ψ_A alone, without transforming ξ_A, produces a projective correspondence of the lines A into each other, without changing the lines B. This results in what may be called a "twisting" motion in the points P. An ordinary geometric rotation requires transformation of A and B together. Thus, an "A transformation" rotates the A line and displaces each point P "upward" in the direction of $+z$ to make a twist. Then a "B transformation" rotates the B lines and displaces the point P "downward" in the direction of $-z$. The net result is that point P is simply "rotated" without displacement but through twice the angle of either the "A transformation" or the "B transformation" alone.

One can perhaps visualize the meaning of the Dirac equation in terms of such notions. Let us consider Schrödinger's suggestion that the electron moves back and forth at the speed of light in a "trembling motion." We can now say that as it is displaced in the $+z$ direction, it turns at the same time, in a net "twisting motion." Then when it returns, in the $-z$ direction, it continues to turn in the same way (so that it has a "twist" of opposite chirality). The final result is not a displacement but a "turning movement" which may be regarded as what is meant by the "spin."

Lastly, it should be noted that one member of a set of conjugate lines, A, can be transformed into a member of the opposite set, B, by a reflection, $x \to -x$. In such a reflection there is effectively an interchange of spinors ψ_A and ξ_A. If one works out the details, ξ_A corresponds to ψ_A in a relationship very similar to the one that is usually called "charge conjugation" in the usual language of relativistic quantum theory. This suggests that the property of charge need not be a primitive notion in terms of the description that we are using here. Rather, it can perhaps (like the light cone) be seen to be yet another formal abstraction from the primitive "spinor" ordering. This possibility will also be discussed in a later paper.

References

1.
D. Bohm, R. Schiller and J. Tiomno, Nuovo Cimento Suppl. (Ser. 10) **1,** 48 (1955); D. Bohm and R. Schiller, Nuovo Cimento Suppl. (Ser. 10) **1,** 67 (1955).

Eight **Helmut Hönl**

A Contribution to the Thermodynamics of the Universe and to 3°K Radiation

1. Formulation of the Problem

A few years ago, radio astronomers established the cosmologically significant result that the universe is filled with isotropic radiation that, according to its spectral energy distribution, should be associated with a temperature of about 3°K.[1–4] The energy distribution appears to correspond with remarkable exactness to Planck's radiation law; also, the total intensity corresponds obviously to radiation belonging to a blackbody of temperature 3°K according to the Stefan-Boltzmann law (hence we are not dealing with "diluted" radiation). The assumption suggests itself that the 3°K radiation represents a kind of relict from the primeval state of the universe, so that its origin must be dated from the time that radiation and matter were still in thermodynamic equilibrium with one another at a very high temperature (perhaps 10^{10}°K or more) and large density. One therefore expects that a study of the behavior of radiation and matter in cosmic expansion would permit, under plausible assumptions, the drawing of some conclusions about the initial state of the universe and about the thermal behavior of radiation and matter during its expansion.

In an article of a nearly similar title, H.-J. Treder[5] has developed the most important results from this conception. Treder's basic assumption—to which we subscribe—is that "from the time of the primeval state of the universe, in which extreme physical conditions reigned, the conversions among the various kinds of energy"—principally rest energy of the matter condensed in the galactic systems, thermal energy of the intergalactic gases (plasma), and radiation energy—"are not cosmologically significant." By contrast, the conversion from rest energy into both radiation and thermal energy still occurs at present in the thermonuclear processes that go on in star interiors. With these idealized, but presumably essentially correct, hypotheses, the behavior of the various kinds of energy during expansion can be investigated independently—to a large degree—of one another on the basis of the Einstein equations for a gravitational field.

The considerations below follow Treder's investigation very closely. They differ, however, from the latter in that he presupposes the validity of the caloric equation of state $P = \frac{2}{3}u$, between the thermal pressure P and the thermal energy density u of the gas (plasma), also for the highest temperatures. But in fact the equation of state holds in this form only for comparatively low temperatures and asymptotically approaches $P = \frac{1}{3}u$ (as in the case of radiation) with increasing temperature. The influence of this circumstance on the discussion of the behavior of plasma as well as of radiation at the expected extremely high temperatures is, indeed, quite considerable. Thus, for example, Treder arrives at the conclusion that with sufficient contraction of the universe (several billion years ago), the plasma temperature T_g must have exceeded the radiation temperature T_s quite significantly, since T_g satisfies an R^{-2} law and T_s an R^{-1} law—R is the world radius, or radius of curvature, of the universe—no matter how small one may estimate the thermal energy of the universe to be in its present stage. Against this contention, it appears more natural to assume that T_g possibly approaches T_s asymptotically with decreasing R or that at a very early stage of the universe, i.e. the primeval state, agreement between T_g and T_s occurred for large energy density and very high temperature. On the other hand, it is probably not permissible to extrapolate T_g and T_s beyond this primeval state to still larger densities (smaller R). In fact, the function $T_g(R)$ for a monatomic relativistic gas (plasma) in its totality turns out to deviate largely from the R^{-2} law; we assume at all times that T_s exceeds T_g.

As concerns the physical nature of the intergalactic gas, we shall have to view the gas as a plasma, that is, as composed essentially of positive and negative charge carriers, mainly protons and electrons (ionized hydrogen).[6] Each type of particle—proton, electron, ion, and neutral atom—then produces its own partial pressure. We shall begin therefore by deriving the gas laws for a uniform monatomic gas over the entire temperature range, while assuming ideal-gas behavior—which is certainly allowable for the plasma—and return to the special discussion of physical conditions only at the conclusion of our treatment.

2. Thermal and Caloric Equations of State of an Ideal Gas at Arbitrarily High Temperature[7]

We mean by "thermal equation of state" of a gas the dependence of pressure on temperature and density, and by "caloric equation of state" the relation between pressure and energy density. Because the temperature can be extremely high, corresponding to the cosmic conditions of the initial state, we must, when deriving these equations, take into account the special-relativistic connection between the energy and momentum of a particle, i.e. $E^2 = (cp)^2 + (m_0c^2)^2$, in place of $E = p^2/2m_0$, m_0 being the rest mass.

Elementary considerations lead to the formula

$$P = \tfrac{1}{3}n\overline{E\beta^2} \tag{8.1}$$

for the pressure of the gas, where E signifies the (relativistic) energy of a particle, $\beta = v/c$ its speed as a fraction of the speed of light, and n the number of particles per unit volume; the average value (denoted by the bar) for a system in thermodynamic equilibrium is to be formed over all n particles in accordance with the Boltzmann distribution law. Since $\beta = cp/E$, one finds, with the notation $\varepsilon = E/m_0c^2$ and on taking into account that $pdp = (m_0c)^2\varepsilon\, d\varepsilon$,

$$\overline{E\beta^2} = \frac{m_0^{-1}\int_0^\infty \varepsilon(p/\varepsilon)^2 e^{-\gamma\varepsilon}p^2\, dp}{\int_0^\infty e^{-\gamma\varepsilon}p^2\, dp}$$

$$= \frac{m_0c^2\int_1^\infty (\varepsilon^2 - 1)^{3/2}e^{-\gamma\varepsilon}\, d\varepsilon}{\int_1^\infty \varepsilon(\varepsilon^2 - 1)^{1/2}e^{-\gamma\varepsilon}\, d\varepsilon}, \tag{8.2}$$

where the abbreviation $\gamma = m_0c^2/kT$ has been used (k is Boltzmann's constant). Integration by parts transforms the numerator of Eq. [8.2] into

$$3\int_1^\infty \varepsilon(\varepsilon^2 - 1)^{1/2}\gamma^{-1}e^{-\gamma\varepsilon}\,d\varepsilon,$$

so that, with reference to [8.2],

$$\overline{E\beta^2} = 3m_0c^2\gamma^{-1} = 3kT. \qquad [8.2']$$

Accordingly, for a confined system of μ moles at the temperature T, Eq. [8.1] leads to the result (note that $knV/\mu = R$, the ideal gas constant)

$$PV = \mu RT \qquad [8.3]$$

as being valid *in all strictness* for arbitrary *high* (as well as ordinary) temperatures. The notion of temperature is, to be sure, tied to the Boltzmann distribution law, whose application presupposes a sufficient degree of interaction (through collisions) among the particles.

We next calculate in a similar manner

$$\bar{E} = \frac{m_0c^2\displaystyle\int_1^\infty \varepsilon^2(\varepsilon^2 - 1)^{1/2}e^{-\gamma\varepsilon}\,d\varepsilon}{\displaystyle\int_1^\infty \varepsilon(\varepsilon^2 - 1)^{1/2}e^{-\gamma\varepsilon}\,d\varepsilon}$$

$$= -m_0c^2\frac{\partial}{\partial\gamma}\ln Z, \qquad [8.4]$$

wherein

$$Z = \int_1^\infty \varepsilon(\varepsilon^2 - 1)^{1/2}e^{-\gamma\varepsilon}\,d\varepsilon.$$

Expanding the integrand in declining powers of ε, one obtains

$$Z = (2\gamma^{-3} + 2\gamma^{-2} + \tfrac{1}{2}\gamma^{-1} + \cdots)e^{-\gamma};$$

and from this, for *high* temperatures,

$$\bar{E} = 3kT + \frac{E_0^2}{2kT} + \cdots, \qquad [8.4a]$$

with $E_0 = m_0c^2$. On the other hand, one finds for *low* temperatures:

$$\bar{E} = E_0 + \tfrac{3}{2}kT + \cdots, \qquad [8.4b]$$

as was to be expected. The molar heat capacity C_v of a monatomic relativistic gas therefore increases continuously from $\frac{3}{2}R$ to $3R$ with increasing temperature (see Fig. 8.1).

On account of the simple form of the thermal equation of state [8.3], it is now an easy matter to arrive at the *caloric* equation of state of the relativistic gas. As the thermal energy density u—including u_0, the density of the rest energy—is given by $u = n\bar{E}$

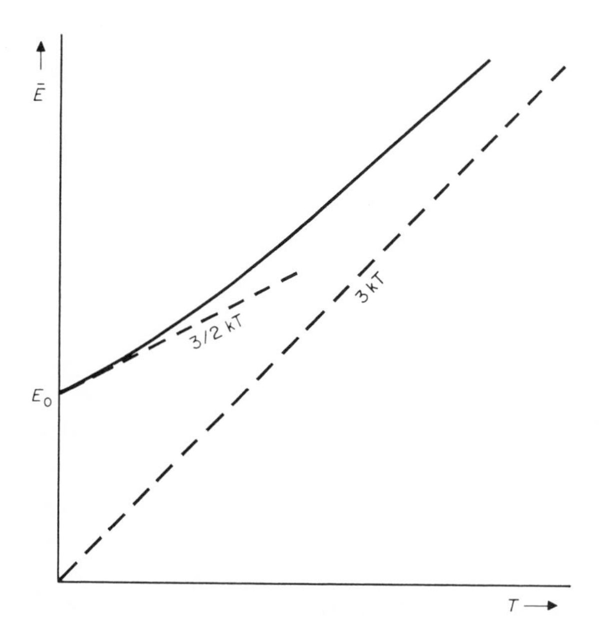

Figure 8.1 Average energy $\bar{E}$ of the monatomic gas as a function of temperature T.

(and likewise, $u_0 = nE_0$) and, on the other hand, $P = nkT$ (from [8.1] and [8.2′]), it follows directly from [8.4a] and [8.4b] that

$$u = 3P + \frac{u_0^2}{2P} + \cdots, \text{ for } u \gg u_0, \qquad [8.5a]$$

and

$$u = u_0 + \tfrac{3}{2}P + \cdots, \text{ for } u - u_0 \ll u_0. \qquad [8.5b]$$

Important for the following, however, is the inverse relation $P = P(u)$; restricting ourselves in each case to the first term of the series, we obtain from [8.5a] and [8.5b]:

$$P = \frac{u}{3} - \frac{u_0^2}{2u} + \cdots, \text{ for } u \gg u_0 \qquad [8.6a]$$

and

$$P = \tfrac{2}{3}(u - u_0) + \cdots, \text{ for } u - u_0 \ll u_0. \qquad [8.6b]$$

The boundary cases can be easily examined. For the highest temperatures, $u \gg u_0$, Eq. [8.6a] furnishes asymptotically $P = \frac{1}{3}u$, as for a photon gas; at low temperatures, $u - u_0 \ll u_0$, we have $P = \frac{2}{3}t$, if $t = u - u_0$ signifies the kinetic energy density, as was to be expected.

Given the boundary results, the transition domain can also be properly surveyed. As the pressure between the boundary situations is without question analytic, one should look for a simple analytic function $P(u)$ that is adapted to the boundary situations and also is convenient for the purpose of the integrations to be carried out later. The simplest procedure for constructing such a function probably consists in augmenting [8.6a] with two further terms, proportional to u^{-2} and u^{-3}, whose coefficients are chosen such that the correct behavior of $P(u)$ according to [8.6b] is assured at the point $u = u_0$ as concerns function value ($P = 0$) and slope

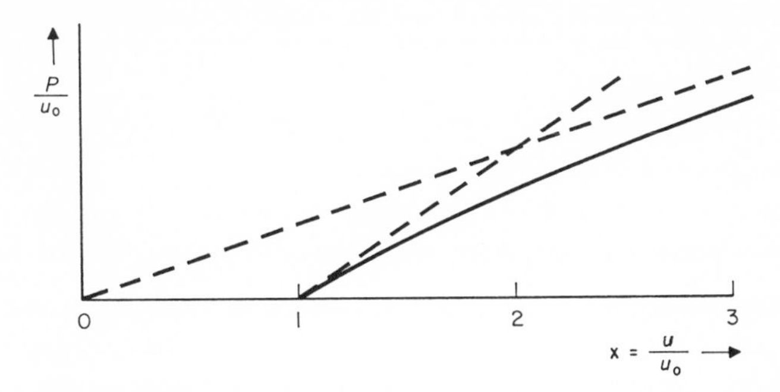

Figure 8.2 Pressure as a function of energy density u, where $x = u/u_0$ is chosen as the abscissa. The slope $\frac{2}{3}$ of the curve at the point $x = 1$ corresponds to the law $P = \frac{2}{3}(u - u_0)$ for low temperatures.

$(dP/du = \frac{2}{3})$. A simple computation thus yields the approximating function

$$P(u) = \frac{u}{3} - \frac{u_0^2}{2u} + \frac{u_0^3}{3u^2} - \frac{u_0^4}{6u^3}. \qquad [8.6]$$

There exists accordingly, for each type of particle, a certain "critical temperature" around which the caloric equation of state for "low" temperatures, [8.6b], gradually goes over into that for a relativistic gas at "high" temperatures, [8.6a]. If we think of the plasma as consisting mainly of free electrons and protons, then we may define the critical temperatures T_e and T_p for the electron gas and proton gas by means of the equations $2m_ec^2 = kT_e$ and $2m_pc^2 = kT_p$, respectively.* The temperatures thus characterized lie at $T_e \sim 10^{11}$°K and $T_p \sim 10^{14}$°K. Starting from normal (very "low") temperatures, the electron gas will therefore become "relativistically degenerate" above 10^{11}°K, the proton gas first above 10^{14}°K; in the intermediate range, between 10^{11} and 10^{14}°K, only the electron gas is relativistically degenerate. The extent to which these temperatures are attained under cosmic circumstances is naturally dependent on the special assumptions concerning the primeval state (see Sec. 5).

* One could also designate the temperatures T_e and T_p as "radiation temperatures" for the electron and proton, respectively. The factor 2 is added arbitrarily in the above definitions.

3. Change of State of Radiation and Gas (Plasma) during Cosmic Expansion

a. Black Radiation and the Conditions under Which It Is Conserved during Cosmic Expansion

We assume that in some particular phase of its development the universe was filled homogeneously and isotropically with "black" radiation of temperature T_s. How does the spectral composition of the radiation change under expansion (or contraction) of the universe?

Black radiation satisfies Wien's law for the distribution of spectral energy:

$$\mathcal{J}_\nu \, d\nu = \nu^3 F\left(\frac{\nu}{T_s}\right) d\nu. \qquad [8.7]$$

Here and in the sequel, we—following A. Friedmann and G. Lemaître—regard the expansion to be a consequence of the Einstein field equations for a metric field; this seems to us the most natural and least hypothetical supposition. Accordingly, the gravitational guidance field of the isotropically distributed matter has, together with the gravitational self-interaction of the radiation, a certain influence on the frequency of the radiation during the expansion. As von Laue[8] first noticed, there exists between the frequency ν and the radius of curvature R of an expanding spherical space* the simple relation

$$\nu R = \text{const.} \qquad [8.8]$$

This means that during the expansion of the universe the frequency will decrease in inverse proportion to R and the frequency interval $d\nu$ will change correspondingly. Let ν' be the altered frequency pertaining to the radius R', then, on account of

$$\frac{\nu}{\nu'} = \frac{R'}{R} \qquad \text{and} \qquad \frac{\nu^3}{\nu'^3}\frac{d\nu}{d\nu'} = \left(\frac{R'}{R}\right)^4,$$

* We base our arguments on the model of a spherical or elliptical space with a positive radius of curvature, because of its graphic quality. In Sec. 4 we shall free ourselves fundamentally from this special conception, but shall keep on favoring the spherical or elliptical model for the reason given.

Eq. [8.7] gives rise to

$$\mathscr{I}_\nu \, d\nu = \left(\frac{R'}{R}\right)^4 \nu'^3 F\left(\frac{\nu' R'}{T_s R}\right) d\nu'. \tag{8.7$'$}$$

That is, we obtain from [8.7] again a Wien spectral energy distribution, but one that belongs to the altered temperature

$$T'_s = \left(\frac{R}{R'}\right) T_s. \tag{8.9}$$

Integration of [8.7′] over the complete spectrum reveals further the connection between the energy densities u_s and u'_s:

$$u_s = \int_0^\infty \mathscr{I}_\nu \, d\nu = \left(\frac{R'}{R}\right)^4 \int_0^\infty \mathscr{I}_{\nu'} \, d\nu' = \left(\frac{R'}{R}\right)^4 u'_s. \tag{8.7$''$}$$

Thus, in addition to

$$T_s R = T'_s R' = \text{const.}, \tag{8.9$'$}$$

the result

$$u_s R^4 = u'_s R'^4 = \text{const.} \tag{8.9$''$}$$

also holds during expansion, in harmony with the energy law of relativity theory (see Sec. 4). Since, by hypothesis, the radiation was originally black, hence $u_s = \sigma T^4$ according to the Stefan-Boltzmann law (σ is the Stefan-Boltzmann constant), we finally get from [8.7″] and [8.9′]:

$$u'_s = \left(\frac{R}{R'}\right)^4 \sigma T_s^4 = \sigma T_s'^4. \tag{8.10}$$

The black nature of the radiation will consequently not be altered by the expansion.

All these phenomena are completely analogous to those that describe black radiation undergoing adiabatic expansion in a container with reflecting walls—as one will easily recognize on substituting the cube root of the volume V for R in Eqs. [8.9′], [8.9″], and [8.10].[9] In this sense we may speak of an "adiabatic cosmic expansion" when referring to the radiation.

For the conjectures regarding the origin of the 3°K radiation the empirical verification of the facts that its energy distribution obeys the Wien-Planck radiation law very precisely and that its total intensity is given by the Stefan-Boltzmann law (with universal constant σ) is now of the greatest importance. By virtue of what was said earlier, one must draw from these data the conclusion that the background radiation of the radio waves emerging from the universe must also have been black during the earlier stages of its evolution. This can, however—if the action of gravitational guidance fields on radiation is assumed—only be interpreted to mean that the present-day 3°K radiation was emitted under conditions of great plasma density and high temperatures, at a time when radiation and matter were still in *mutual thermodynamic equilibrium.*

Attention should here be directed to one point in particular: The preceding considerations are obviously tied only to the two assumptions [8.7] and [8.8]. One may ask what will change in these considerations if one replaces the constant gravitation coefficient κ (as presumed by the Einstein gravitational field equations) in the computations by a variable κ (as an additional field variable) in the sense of the Dirac-Jordan hypothesis. One can hardly doubt the justification of the assumption [8.7]; for, at sufficiently large density of matter in the primeval state, the appearance of thermodynamic equilibrium, and hence the Planck-Wien spectral distribution, will surely follow within a time interval that is vanishingly small compared with the times characterizing changes of state as a result of cosmic expansion. Against this thesis, the correctness of posit [8.8] for a variable coefficient κ must first be proved (by a suitable modification of von Laue's proof). Such a proof, for which I am indebted to my collaborator H. Dehnen, may now indeed be given in a direct way for a homogeneous and isotropic expanding universe.* Hence the laws [8.9′], [8.9″], and [8.10] follow also in the general case of a time-dependent κ from Eqs. [8.7] and [8.8]. On the other hand, from the field equations of a gravitational field

* This will be commented upon at another place [Z. Physik **68,** 190 (1968)].

for the expansion of a homogeneous and isotropic universe with variable κ (generalized Friedmann solution), P. Jordan[10] derived the conservation equation

$$\kappa^2 u_s R^4 = \text{const.}, \tag{8.9a}$$

in contrast to Eq. [8.9″]. Equation [8.9a] expresses, however, a different behavior of the radiation energy density u_s under expansion than Eq. [8.9″], for [8.9′] and [8.9a] now yield

$$u_s' = \left(\frac{\kappa}{\kappa'}\right)^2 \sigma T_s'^4 \tag{8.10a}$$

in place of Eq. [8.10]. By starting from Jordan's law [8.9a] instead of from [8.9″], one must conclude that the (empirically established) black 3°K radiation was emitted in the primeval state at a much larger κ than "diluted" black radiation—a result that is unacceptable thermodynamically. From the only thermodynamic assumption possible, namely, that the cosmic background radiation was originally emitted as black radiation (i.e. that it possessed in the beginning the Planck-Wien spectral energy distribution), it would follow conversely from [8.10a] that the radiation evident today must be "condensed"—a result contradicting observation.* The solution of the contradiction obtains only for κ = const. In the fact that the observed 3°K radiation still today, i.e. supposedly several billion years after its emission, has the character of non-condensed black radiation, we may see an important empirical

* That the cosmic background radiation is neither condensed nor diluted is shown, within quite fine tolerances, by observation. It should here be remarked above all that the determination of the radiation *temperature* at *one* of the four wavelengths employed, viz. at λ = 0.26 cm, could be carried out as an "indirect" measurement according to the CN method; this measurement makes it possible to establish a calibration point for the Planck energy distribution function. The *intensity* measurements at the three wavelengths λ = 3.2 cm, 7.35 cm, and 20.7 cm lie rather exactly on the 2.7°K isotherm. Since these three wavelengths for about T = 3°K still lie completely within the Rayleigh-Jeans spectral region, the intensity measurements alone would not have enabled us to draw any definite conclusion about the radiation temperature, as they permit a determination of the radiation energy density only when combined with the CN method (cf. Refs. 1–4).

argument in favor of a *constant* gravitation coefficient κ, or in any event a κ that can vary only weakly!

b. Monatomic Gas (Plasma) Behavior in Cosmic Expansion

It is of interest to examine whether a monatomic gas (plasma) has a behavior analogous to that of radiation under cosmic expansion. One must distinguish here between low, medium, and very high temperatures.

Beginning with low temperatures, let us suppose that the Maxwell-Boltzmann distribution law

$$w(p, T) = Ae^{-p^2/2m_0kT}p^2\,dp \qquad [8.11]$$

holds in a particular stage of development (at "large" R). Is this distribution maintained during cosmic expansion (or contraction)?

One can easily show that during the change of the radius of a homogeneous and isotropic spherical space filled with matter (and radiation) the following relation (analogous to Eq. [8.8]) holds rigorously,[11] according to the Einstein theory, for the momentum p:

$$pR = \text{const.}^* \qquad [8.12]$$

Hence, in the case of isotropic cosmic expansion (or contraction) $R \to R'$, Eq. [8.11] becomes

$$w(p', T') = A'e^{-(p'^2/2m_0kT)(R/R')^2}p'^2\,dp'. \qquad [8.11']$$

We thus again get a Boltzmann distribution, but one belonging to the temperature

$$T' = \left(\frac{R}{R'}\right)^2 T \qquad [8.13]$$

so that, during isotropic cosmic expansion,

$$TR^2 = T'R'^2 = \text{const.} \qquad [8.13']$$

* Eq. [8.8], $\nu R = \text{const.}$, is a special case of [8.12], as one recognizes on replacing p by the momentum $h\nu/c$ of a photon.

The change in the kinetic energy density t is found most simply from $\bar{E}_{\mathrm{kin}} \propto T$, valid per particle for sufficiently low temperatures; see [8.4b]. Since the number of particles remains unchanged during expansion, the particle density $n \propto R^{-3}$ decreases in this process. Therefore, by virtue of [8.13′],

$$t = n\bar{E}_{\mathrm{kin}} \propto R^{-5},$$

that is,

$$tR^5 = \text{const.} \qquad [8.14]$$

If R is replaced by $V^{1/3}$ in [8.13′] and [8.14], one obtains again the laws for adiabatic expansion of monatomic gases in containers.

These conclusions cannot, however, be carried over to the medium temperature range where, because of the relativistic energy-momentum relation, $\bar{E}_{\mathrm{kin}}$ can no longer be set proportional to T; see [8.4a] and [8.4b]. It thus appears that the Boltzmann distribution for medium and high temperatures,

$$w(p, T) = Ae^{-c(p^2+m_0^2c^2)^{\frac{1}{2}}/kT}p^2\,dp,$$

is not reproduced under a transformation of the momenta defined by [8.12]. If, therefore, we assume, as heretofore, that no interaction (through collisions) occurs among the particles in the course of the expansion, then the concept of temperature becomes illusory for the expanding cosmic gas. The assumption of such an interaction does not, to be sure, contradict the hypothesis that from the time of the primeval state essentially no further interaction between gas and matter should have occurred. For this reason, when we continue to speak of the "temperature" of the gas for the whole intermediate region between very high and low temperatures, we want to imagine that, during the long cosmic time intervals, sufficiently many collisions have taken place, so that the equilibrium state (maximum of entropy) could appear at *each* stage of development. This supposition must always be understood in the sequel; it is important for the temperature definition, since only by assuming a

Boltzmann distribution may we specify a definite temperature of the gas. At extremely high temperatures the conditions of the gas approach more and more those of the radiation, as one may easily see, so that in this range of highest energies the concept of temperature again becomes unequivocal, even without the assumption of interaction among the particles.

4. Integration of the Friedmann-Lemaître Energy Equation for the Expanding Universe

In the foregoing we have shown that in an expanding universe the guidance field exercises a decisive influence on the changes of state of radiation and plasma. The preceding results can consequently also be established through integration of the Einstein field equations.* The "energy equation" derived from these by A. Friedmann and G. Lemaître now, however, also allows us to bridge the gap between extremely high and low temperatures for a monatomic gas (plasma), a goal that could be reached only with difficulty along the way of statistical thermodynamic reasoning.

According to Friedmann and Lemaître, the expansion of a homogeneous and isotropic universe is governed by the two differential equations

$$\dot{U} + 3\left(\frac{\dot{R}}{R}\right)(U + P) = 0 \qquad [8.17]$$

and

$$R^{-2}(2R\ddot{R} + \dot{R}^2 + \varepsilon c^2) = -\kappa c^2 P. \qquad [8.18]$$

Herein R represents the linear measure of curvature, U the total energy density, and P the total pressure; $\varepsilon = \pm 1, 0$ corresponds respectively to positive, negative, and vanishing space curvature.

* H.-J. Treder took this route in the work cited. Our subsequent considerations consequently parallel to a great extent those of Treder. For the reader's convenience some repetition may perhaps be permitted here.

The connection among energy density, pressure, and expansion is already revealed by the "energy equation" [8.17] alone.* Equation [8.18] describes the temporal progress of the expansion. This will not be pursued any further here; we remark only that $R(t)$, starting out from the value zero, grows rapidly at first but later at a rate that continuously declines (Hubble constant $\propto \dot{R}/R$).

On the other hand, $U(R)$ and $P(R)$ are, by Eq. [8.17], completely independent of the parameter ε, signifying the curvature of space. For graphic purposes as well as physical reasons, we shall put ε equal to unity in the following and hence assume a closed finite universe. (A bounded volume increases proportionally to R^3 also in the case of an arbitrary curvature sign.)

We describe the properties of state of the universe in this manner:
1. Let the rest energy density residing in the galaxies be $U_0 = \rho c^2 \propto R^{-3}$; the kinetic energy of the galaxy movements, in contrast, should be neglected—its density being very much less than $10^{-35}c^2$ gm cm^{-3},† whereas $\rho \sim 10^{-28}$ to 10^{-30} gm cm^{-3}.
2. Let the average energy density of the gas (plasma) in the intergalactic space be denoted, as before, by u; it is the sum of the rest energy density u_0 and the kinetic energy density t, whence $t = u - u_0$. For the gas pressure we use the symbol p (rather than P as in Secs. 2 and 3). The determination of $u(R)$ and $p(R)$, and from this the gas temperature $T_g(R)$, will be our main task in the following paragraphs.
3. Let, finally, the radiation energy density again be u_s and the radiation pressure $p_s = u_s/3$. In the course of integration, the radiation temperature T_s and u_s will prove to be in agreement with Eqs. [8.9′] and [8.9″].

* The equation $d(UV) + PdV = 0$ follows directly from [8.17] if $V \propto R^3$ is a somehow bounded, coexpanding volume. Since $d(UV) + PdV = TdS$ is the reversible work performed during a change of volume dV, one can view [8.17] as an expression of the constancy of the entropy S under cosmic expansion ($dS = 0$). Hence Eq. [8.17] could be more properly called the *entropy law.*

† From personal correspondence with H.-J. Treder.

The total energy density U and the total pressure P are, accordingly, composed as follows:

$$U = U_0 + u + u_s,$$

and

$$P = 0 + p(u) + \tfrac{1}{3}u_s. \qquad [8.18]$$

Under the simplified assumption that during the expansion of the universe from its primeval state no essential further interaction of the three kinds of energy has occurred (see Sec. 1), the energy equation [8.17] may be resolved into these three equations:

$$\dot{U}_0 + 3\left(\frac{\dot{R}}{R}\right) U_0 = 0, \qquad [8.19a]$$

$$\dot{u} + 3\left(\frac{\dot{R}}{R}\right)(u + p) = 0, \qquad [8.19b]$$

$$\dot{u}_s + 4\left(\frac{\dot{R}}{R}\right) u_s = 0. \qquad [8.19c]$$

Equations [8.19a] and [8.19c] can be integrated immediately to yield

$$U_0R^3 = \text{const.}, \qquad \text{or} \qquad U_0V = M_0c^2 = \text{const.}, \qquad [8.20a]$$

and

$$u_sR^4 = \text{const.} \qquad [8.20b]$$

Equation [8.20a] expresses the conservation of the total rest mass of the universe—to be supplemented indeed by u_0V after subtraction of the rest energy of the gas (see below). Equation [8.20b] is in agreement with [8.9″]. From this, together with [8.10], it follows that

$$T_sR = \text{const.} \qquad [8.21]$$

Our most important further task is the integration of [8.19b]. On separation of u_0 from u, this equation decomposes into

$$\dot{u}_0 + 3\left(\frac{\dot{R}}{R}\right) u_0 = 0$$

and

$$\dot{t} + 3\left(\frac{\dot{R}}{R}\right)(t + p) = 0. \qquad [8.22]$$

If additionally we again consider the boundary cases and correspondingly put $p = \frac{1}{3}t$ for extremely high, and $p = \frac{2}{3}t$ for extremely low, temperatures, then Eq. [8.22] furnishes:
on the one hand,

$$u_0 R^3 = \text{const.}, \qquad [8.23]$$

analogous to [8.20a]; on the other,

$$tR^4 \approx uR^4 = \text{const., for extremely high temperatures,} \qquad [8.24a]$$

and

$$tR^5 = \text{const., for low temperatures.} \qquad [8.24b]$$

We note that [8.24a] and [8.24b] agree with [8.9″] and [8.14], respectively.

In the general case, i.e. for the intermediate range between low and extremely high temperatures, one must replace p in [8.19b] with the function $p(u)$ corresponding to our approximation [8.6] (with p instead of P). It will suffice here to limit ourselves to the first two terms of [8.6], especially since the remaining terms (added only for the sake of the joining at $u = u_0$) are uncertain. If, by virtue of [8.23], we set $u_0 \propto R^{-3}$, the Eqs. [8.19b] and [8.6] (or [8.6a]), give rise to this differential equation for $u(R)$:

$$\dot{u} + \frac{\dot{R}}{R}\left(4u - \frac{C^2}{uR^6}\right) = 0. \qquad [8.25]$$

When integrated, this yields

$$u(R) = R^{-4}(C^2R^2 + b^2)^{1/2}, \tag{8.26}$$

with b as a new integration constant, while C, by [8.23], is essentially the total mass of the gas. The function $u(R)$ is represented in a very satisfactory manner by the formula [8.26]: For a state of highly concentrated matter (small R) and correspondingly high temperature, [8.26] reduces to $u \approx bR^{-4}$, in accord with [8.24a]; whereas for strong dilution of the gas (large R) and low temperatures, [8.26] gives $u \approx CR^{-3}$, i.e. a value essentially equal to u_0.*

With u known as a function of R, one gets from [8.6] the pressure p as a function of R. Introduction of [8.26] into [8.6] leads, if we furthermore set $u_0 = CR^{-3}$,* to the result

$$\begin{aligned} p(R) = {} & \tfrac{1}{3}R^{-4}(C^2R^2 + b^2)^{1/2} - \tfrac{1}{2}R^{-2}C^2(C^2R^2 + b^2)^{-1/2} \\ & + \tfrac{1}{3}R^{-1}C^3(C^2R^2 + b^2)^{-1} - \tfrac{1}{6}C^4(C^2R^2 + b^2)^{-3/2}. \end{aligned} \tag{8.27}$$

In what follows it is convenient to use instead of R the variables z and x defined as

$$z = \frac{C}{b} R \qquad \text{and} \qquad x = z^{-1}(z^2 + 1)^{1/2} \; (\geq 1). \tag{8.28}$$

This allows us, first of all, to write u and p according to [8.26] and [8.27] in the simplified forms:

$$u(z) = \frac{C^4}{b^3} \frac{(z^2 + 1)^{1/2}}{z^4} \tag{8.26$'$}$$

and

$$\begin{aligned} p(z) = \frac{C^4}{b^3} [& \tfrac{1}{3}z^{-4}(z^2 + 1)^{1/2} - \tfrac{1}{2}z^{-2}(z^2 + 1)^{-1/2} \\ & + \tfrac{1}{3}z^{-1}(z^2 + 1)^{-1} - \tfrac{1}{6}(z^2 + 1)^{-3/2}]. \end{aligned} \tag{8.27$'$}$$

* One would expect that, for large R, u goes asymptotically over into u_0. If we set $u_0 = C'R^{-3}$, then the relation of C in [8.25] and [8.26] to C' is $C = (\frac{4}{3})^{1/2}C'$. This trivial difference between C and C' is related to our discarding the higher terms of the series [8.6] when substituting it into [8.19b].

With the use of these equations, the thermal equation of state [8.3], $pV \propto T$, provides the gas temperature T_g. If we note that $V \propto z^3$, then T_g as a function of x is written for the present with an undetermined coefficient:

$$T_g = Af(x), \qquad [8.29]$$

where

$$f(x) = \tfrac{1}{3}x - \tfrac{1}{2}x^{-1} + \tfrac{1}{3}x^{-2} - \tfrac{1}{6}x^{-3}. \qquad [8.29a]$$

We recognize $f(x)$ as the function $p(u)/u_0$ in Eq. [8.6]—now with $p(u)$ written instead of $P(u)$—provided we set $x = u/u_0$ (as in Fig. 8.2). If, for the purpose of comparison, the radiation temperature $T_s \propto R^{-1}$ is also expressed in terms of the variable x instead of R, one gets

$$T_s = B(x^2 - 1)^{1/2}. \qquad [8.30]$$

The constants A and B are disposable and can be parametrized to the selected particular physical stipulations.

5. Discussion of the Results

According to the sketch we have given at the outset of the origin of the 3°K radiation, it should be assumed that $T_s \approx T_g$ in the so-called primeval state; and at that time the present-day 3°K radiation should have originated as black radiation at a comparatively very high temperature and a great density. Beginning with $T_s \approx T_g$, the temperatures of the radiation and the plasma have since then decreased during expansion according to various laws, as is expressed by Eqs. [8.29] and [8.30]. It is, in fact, easy to gather from these equations that, in the course of the expansion following the primeval state, T_s must always have been larger than T_g and that the ratio T_s/T_g assumed more and more extreme values with increasing R.

A more reliable, careful discussion of the behavior of T_s and T_g during expansion is now greatly facilitated by the use of the variable

x in place of R. In order to pursue this interpretation in better detail, let us plot, in the same diagram, both T_s and T_g as functions of x according to [8.29], [8.29a], and [8.30]. It is convenient for this purpose to consider, first of all, the boundary case where asymptotic agreement of T_s and T_g occurs for very small R, i.e. practically at $R = 0$, or as $x \to \infty$. For this case, the factors A and B are to be chosen such that the curves $Af(x)$ and $B(x^2 - 1)^{1/2}$ touch asymptotically as x approaches infinity. This is realized for $A/B = 3$. Let us therefore take $A = 1$ and $B = \frac{1}{3}$ in our diagram—to obtain the two curves depicted in Fig. 8.3. The curve for T_s is here a branch of a hyperbola, which the curve $T_g = f(x)$ approaches

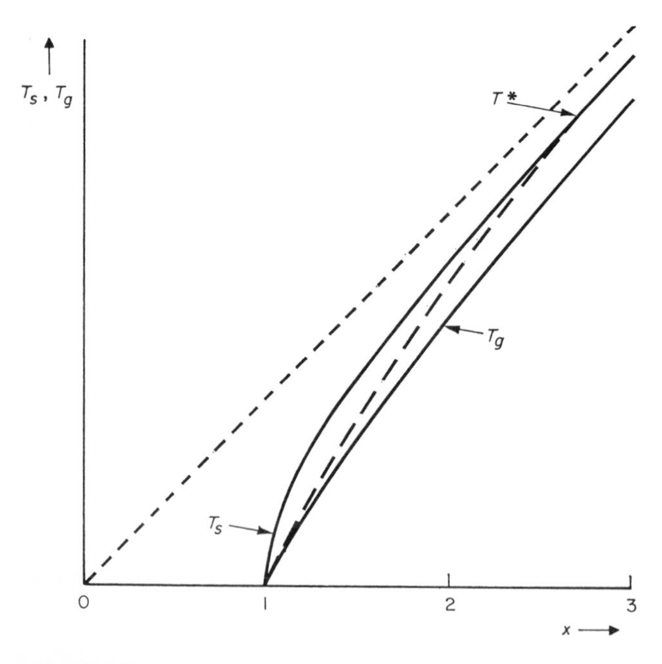

Figure 8.3 Radiation and gas temperatures, T_s and T_g, as functions of the cosmic expansion, with abscissa x as defined in Eq. [8.28]. The full curves relate to the boundary case of extremely high initial temperature and density values; the broken curve represents the variation of T_g on assuming a smaller (but still very high) initial temperature T^*.

asymptotically from below with increasing x (i.e. decreasing R); at the point $x = 1$ ($R \to \infty$), the curves meet as T_s and T_g both become zero. At all other points, T_s is obviously always larger than T_g.

The relation between these two functions in the neighborhood of $x = 1$, that is, for low temperatures, is because of [8.29], [8.29a], and [8.30] given by

$$T_g \approx [2(x - 1)]^{1/2} T_s. \qquad [8.32]$$

Thus, the larger R is, the more T_s predominates over T_g, corresponding to their behavior at low temperatures: $T_s \propto R^{-1}$ and $T_g \propto R^{-2}$. Because $x = u/u_0$, Eq. [8.32] may also be written as

$$T_g \sim \left(\frac{t}{\rho_1 c^2}\right)^{1/2} T_s, \qquad [8.32']$$

where t is the kinetic energy density and $\rho_1 c^2 = u_0$ the rest energy density of the gas. At first little is known about these two quantities, but one may nevertheless assume that $t \ll \rho_1 c^2$ in any event.

For the limiting case of very high initial temperature that we are aiming at, the following situation appears to hold convincingly. Let us imagine the gas, or plasma, to consist of protons and electrons —at high temperatures a neutral mixture, at low temperatures H atoms. Then $t \sim nkT_g$ if $n = \rho_1/m_p$ stands for the number of protons per unit volume; m_p is the proton mass, compared with which the electron mass m_e may be neglected. We thus find

$$T_g \sim \frac{m_p t}{\rho_1 k}. \qquad [8.33]$$

Elimination of T_g between [8.32′] and [8.33] leads for sufficiently low temperatures, to

$$\left(\frac{t}{\rho_1 c^2}\right)^{1/2} \sim \frac{kT_s}{m_p c^2} \sim \frac{T_s}{T_p}.$$

(We recall that $T_p \sim 10^{14}$°K is the radiation temperature of the proton; see Sec. 2.) On substituting this expression in [8.32′], one observes that

$$T_g \sim \frac{T_s^2}{T_p} \sim 3 \times 10^{-13}\text{°K}. \qquad [8.34]$$

Analogously, it is found that

$$t \sim \left(\frac{T_s}{T_p}\right)^2 \rho_1 c^2 < 10^{-34} \text{ erg cm}^{-3}, \qquad [8.35]$$

if one assumes $\rho_1 < \rho$, where $\rho \sim 10^{-30}$ to 10^{-28} gm cm^{-3} signifies the astronomically estimated average mass density of the universe. Noteworthy is the fact that all special suppositions—for example, about the average mass density of the universe—have disappeared from Eq. [8.34]; it was assumed only that the mass density of the plasma agrees essentially with that of the protons.

To be sure, these extreme circumstances correspond only to the boundary case $A/B = 3$—the curves for T_s and T_g touch in Fig. 8.3 as $x \to \infty (R \to 0)$, corresponding to an extremely high initial temperature $T^* > T_p$. Actually the relation $A/B \geq 3$ is unknown. If the numerical value of A/B is left undetermined, we obtain

$$T_g \sim \left(\frac{A}{3B}\right)^2 \frac{T_s^2}{T_p} \qquad [8.34']$$

and

$$t \sim \left(\frac{A}{3B}\right)^2 \left(\frac{T_s}{T_p}\right)^2 \rho_1 c^2, \qquad [8.35']$$

in place of [8.34] and [8.35]. From [8.34′] one infers an initial temperature for $T_g = T_s$ as

$$T^* \sim \left(\frac{3B}{A}\right)^2 T_p \leq 10^{14}\text{°K}. \qquad [8.36]$$

The ratio A/B left undetermined in Eqs. [8.34′], [8.35′], and [8.36] can, on principle, be ascertained from the first two in case

T_g and T_s, or t and u_s, are accessible to measurement. Both quantities T_g and t are, however, imperceptibly small.* We consequently lack here a decisive datum from which, along with T_s (and u_s), the primeval state of the universe—its initial temperature and density—may be inferred. At this point, therefore, occasion exists for special hypotheses, into which, however, we shall not enter here.[12]

We should still make special mention of the fact that also arbitrarily high initial temperatures and correspondingly large densities ("Big Bang" theory) are compatible with our considerations. If, by chance, $T^* \sim T_p \sim 10^{14}$°K, then the initial mass density ρ^* must have been larger than the present mass density of the universe, $\rho \sim 10^{-30}$ to 10^{-28} gm cm^{-3}, by a factor of about $(10^{14})^3 = 10^{42}$, i.e. $\rho^* \sim 10^{12}$ to 10^{14} gm cm^{-3}—a density comparable, in order of magnitude, with that of nuclear fluid. At such large densities, we indeed would have to replace the thermodynamic equation of state for ideal monatomic gases by the equation of state of a Fermi gas for neutrons, which matter we shall not discuss any further here.

In conclusion, our results, admittedly incomplete, may be derived on the basis of a few assumptions, to wit, that: (1) the positive expansion of the universe is a consequence of Einstein's equations for the guidance field (constant gravitational coefficient κ); (2) the

* The energy density of the primeval heat still in existence can be estimated in the following manner. Let $t \sim nkT_g = (\rho_1/m_p)kT_g = \lambda \rho c^2$, where ρ_1 is the gas density, ρ the average density of matter in the universe, and λ the sought proportionality factor $t/\rho c^2$. Since $\rho_1 < \rho$, we have $\lambda \sim \rho_1 T_g/\rho T_p \leq T_g/T_p$. On the other hand, we have $T_g \ll T_s = 3$°K; consequently $\lambda \ll T_s/T_p \sim 10^{-13}$. Now, according to an estimate given in G. McVittie, *General Relativity and Cosmology*, 2d ed. (Wiley, 1956), p. 188, the total density of all star and star-system energies —kinetic energy of the galaxies and stars, as well as the internal energy of the stars, including radiation produced through local processes—amounts to *at most* a millionth part of the rest energy density ρc^2. Thus the thermal energy density is still smaller than that by a factor much less than 10^{-7}; it is therefore without relevance, from an astronomical standpoint at least.

plasma in the primeval state consisted predominantly of free protons and electrons; and, finally, (3) thermodynamic equilibrium existed initially between radiation and plasma. From this primeval state, the universe should have developed with increasing expansion up to the present state, wherein, apart from the general flight of the galaxies, the 3°K radiation above all bears witness to the earlier state of affairs.

[Just before this article went to press, the author added the following remarks.—The Editors]

The foregoing considerations need to be supplemented by two points.

1. Briefly, after the completion of this article, P. Jordan was able to demonstrate, in an article in *Z. Astrophys.*, that my remarks about 3°K radiation (Sec. 3a) are compatible with Dirac's hypothesis of a variable gravitational constant κ, at least in principle.[13] In consequence, however, one has to forgo the special formulation of this hypothesis as it was presented by Jordan in his book *Schwerkraft und Weltall*.[14] In this case, the modified theory of Jordan then becomes identical with the scalar-tensor theory of the gravitational field developed by C. Brans and R. H. Dicke.[15] In their model the conservation laws retain the same form as in Einstein's original theory. Thereby we are able to exclude an increase of the number of light quanta in connection with any cosmic expansion. My criticism of a variable gravitational constant referred to the above-mentioned interpretation and development of Dirac's hypothesis by Jordan and becomes obsolete in view of the new formulation of the theory (cf. the conclusion of Sec. 3a).

2. Views about the various states of the universe preceding the epoch—similar to the present situation—when the cosmos consisted essentially of photons, electrons, and protons have been avoided in this article for fundamental reasons. All speculations that go beyond this, and these have been frequent during the past few years, seem to presuppose that the supposition of homogeneity

and isotropy is just as justified for the early stages of the universe as for the current state of the cosmos. I do not think that this assumption can be seriously defended. To the contrary, there appears to exist a very strong argument against such a supposition when one considers the rather considerable *Eigenbewegungen* of the galaxies with an average velocity of some hundred km/sec. On extrapolation of these velocities to the early stages of the universe—according to the law of pR = const. that holds for galaxies also (Eq. [8.12])—we obtained the result that the velocity of the matter which now forms our galaxies must have approximated the velocity of light throughout. This means that the whole universe must have been in a state of very turbulent convective motion of matter, and this contradicts the assumption of matter at rest (relative to the substratum) and equally distributed. Conversely, it is also not easy to understand how the current state of the universe with its still considerable *Eigenbewegungen* of the galaxies could have originated from an initially homogeneous and isotropic universe. For this reason we must regard, at least for the time being, all speculations as to the earliest states of the universe (e.g. the origin of the fundamental particles and of the elements) with utter skepticism.

References

1.
A. A. Penzias and R. W. Wilson, Astrophys. J. **142**, 419 (1965); see also R. H. Dicke, P. J. E. Peebles, P. G. Roll, and D. T. Wilkinson, Astrophys. J. **142**, 414 (1965).
2.
P. J. E. Peebles, Phys. Rev. Letters **16**, 410 (1966).
3.
T. F. Howell and J. R. Shakeshaft, Nature **210**, 1318 (1966).
4.
P. G. Roll and D. T. Wilkinson, Ann. Physik **44**, 289 (1967).
5.
H.-J. Treder, Forsch. und Fortschr. **41**, 132 (1967).
6.
Ref. 1.
7.
Cf. J. L. Synge, *The Relativistic Gas* (North-Holland, 1957); R. C. Tolman

Relativity, Thermodynamics, and Cosmology (Clarendon Press, 1934), pp. 118 ff. and 291 ff.
8.
M. von Laue, Berl. Ber. 1931, p. 123; see also M. von Laue, *Die Relativitätstheorie* (Vieweg, 1956), vol. II, §52c.
9.
M. Planck, *Theorie der Wärme* (Hirzel, 1930), §§108 and 109.
10.
P. Jordan, *Schwerkraft und Weltall*, 2d ed. (Vieweg, 1955), bk. 2, p. 189.
11.
H. Hönl, Ann. Physik **6**, 169 (1949).
12.
See, for example, H. Alfvén, *Kosmologie und Antimaterie* (Umschau-Verlag, 1967).
13.
P. Jordan *Z. Astrophys.* **68**, 201 (1968); see also *Z. Astrophys.* **68**, 181 (1968).
14.
Ref. 10.
15.
C. Brans and R. H. Dicke, Phys. Rev. **124**, 925 (1961).

Nine

Dennis Caldwell
and Henry Eyring

Quantum-Mechanical Rate Processes

Introduction

The authors, like so many others, are indebted to Professor Landé for his illuminating contributions to atomic structure. For this reason we deeply appreciated being invited to contribute to this volume in his honor.

We first use the time-dependent Schrödinger equation to treat rate processes. An exact procedure would lead to serious difficulties, but it is possible to formulate the problem and so get a feeling for the nature of the difficulties.

Having done this, we develop the equilibrium theory of reaction rates which has been successfully applied in a wide variety of cases. This involves treating the system at the transition state as an activated complex, like other molecules, except that it has a fourth translation corresponding to motion along the reaction coordinate. The equilibrium rate of crossing the barrier must then be corrected for new equilibrium conditions and for quantum effects. This correction is formally lumped into a mean.

Time-Dependent Quantum Processes

The treatment of complex molecular processes has often been greatly simplified by the division of motion into classical and nonclassical modes. In the equilibrium situations encountered in statistical mechanics, time has been eliminated as a parameter; yet the most intimate details of the motion are decidedly time dependent. Fortunately only certain average quantities are required to describe the properties of matter in the equilibrium state. The two extremes of quantum and classical motion are generally encountered together when most systems are well above absolute zero. For example, at 273°K and 1 atm pressure the atoms of helium gas will be mostly in the ground quantum state, while the translational motion can be regarded as classical. The criterion lies in the ratio of the quantum energies to kT. The lowest levels of the helium atom and their spacing are quite large, whereas the spacing of levels for a free particle in a reasonably sized volume is relatively small.

These arguments apply strictly to rarefied systems in which the average collision time is small compared with the time between collisions. During a collision the rate of change of a relative coordinate decreases to zero and then changes sign. In this region the motion is nonclassical; however, if the density is low, most of the motion will be classical and the quantum corrections are small.

In a crystal the electrons are arranged in quantum levels which provide a field to control the thermal motion of the nuclei. Again at high temperatures the motion is classical and at absolute zero the nuclear motion is described by standing waves of a purely quantum nature. This picture is corroborated by the behavior of the specific heat, which for many crystals approaches the classical value of $3R$ cal/deg/mole at high temperatures. It has been shown that the probability distribution for the coordinates of a system of oscillators in thermal equilibrium is

$$\rho(Q) = \left(\frac{\omega}{\pi\hbar} \tanh \frac{\hbar\omega}{2kT}\right)^{1/2} \exp\left[-Q^2 \frac{\omega}{h} \tanh \frac{\hbar\omega}{2kT}\right]. \qquad [9.1]$$

At high temperatures one may use the approximation

$$\tanh \frac{\hbar\omega}{2kT} \sim \frac{\hbar\omega}{2kT}$$

and

$$\rho(Q) \sim \frac{\omega}{\sqrt{2\pi kT}} \exp\left[-\frac{\omega^2 Q^2}{2kT}\right],$$

which is the classical probability for finding a system with energy $\frac{1}{2}\omega^2 Q^2$. It is computed under the assumption that all regions of phase space are equally probable. At low temperatures

$$\tanh \frac{h\omega}{2kT} \to 1$$

and

$$\rho(Q) \sim \sqrt{\frac{\omega}{\pi h}} \exp\left[-\frac{\omega Q^2}{\hbar}\right],$$

which is the quantum probability distribution for an oscillator in its ground state. When it comes time to discuss the phenomenon of molecular decomposition it will be instructive to recall this simple example.

When the refuge of equilibrium is relinquished it becomes necessary to introduce time in some specific way. One of the most remarkable features in the classical theory of transport phenomena is the derivation of differential equations which are first order in time, while the substratum of particles obeys a system of second-order equations. This is also true in reaction kinetics.

It is evident that in most time-dependent processes of interest a strong statistical correlation of the molecules is still maintained. For example, in considering the conductivity of a gas the temperature of the boundary is fixed and the steady-state temperature is found from an equation of continuity. The amount of energy transported from one part of the boundary to another is determined by the value of ∇T. In the simple case of two infinite parallel planes at different temperatures the energy flow per unit area is $\kappa(dT/dx)$ evaluated at either one of the boundaries.

A summary of the elementary analysis of the problem is worthy of inclusion.

1. The medium is assumed to have a continuous Boltzmann distribution of velocities, $e^{-p^2/2mkT}$, where T is the local temperature.
2. The flow of gas molecules across an arbitrary plane is assumed equal in both directions, to first order. The average value of the speed along a particular direction is found to be one-half the average value of the so-called peculiar speed $\bar{c}$. The number of molecules crossing per unit area in either direction is then given by $\frac{1}{4}\rho\bar{c}$, where ρ is the number density.

3. Molecules crossing from left to right are assumed to have the energy associated with a layer one mean free path λ to the left of the reference plane, with a similar assumption about those traveling in the opposite direction. To first order, the net energy transferred per unit area per unit time is

$$(\tfrac{1}{4}\rho\bar{c})\left(-2\lambda\frac{\partial E}{\partial x}\right).$$

Since

$$\frac{\partial E}{\partial x} = \frac{\partial E}{\partial T}\frac{\partial T}{\partial x} = c_v\frac{\partial T}{\partial x},$$

the final formula for the energy flow is

$$\mathbf{J}_E = -\kappa\,\nabla T,$$

where

$$\kappa = \tfrac{1}{2}\rho\bar{c}\lambda c_v.$$

4. The macroscopic behavior of a fluid subject to temperature or concentration gradients is determined by a continuity equation expressing conservation of energy or mass. For conductivity problems it takes the form

$$\frac{\partial\rho_E}{\partial t} + \nabla\cdot\mathbf{J}_E = 0,$$

which is nothing more than the statement that rate of increase of energy in a given volume is equal to the net flow across the bounding surface. Since

$$\mathbf{J}_E = -\tfrac{1}{2}\rho\bar{c}\lambda c_v\,\nabla T$$

and

$$\rho_E = \rho[c_v T + \text{const.}],$$

the continuity equation takes the form

$$\frac{\partial T}{\partial t} = \nabla(\sigma\,\nabla T),$$

where

$$\sigma = \tfrac{1}{2}\bar{c}\lambda$$

is the coefficient of thermal diffusion.

We have thus progressed from a microscopic many-body system governed by a system of second-order differential equations in time to a first-order equation in time. (The Hamilton-Jacobi equation in classical mechanics is a system of second-order ordinary differential equations in disguise; whereas the thermal diffusion equation is truly first order in time for a function of only four variables, $T(x, y, z, t)$.)

Time now enters as a parameter correlating the interaction of a quasi continuum of volume elements, each of which is in a slightly different state of near statistical equilibrium. As seen from the preceding arguments, the time it takes for these elements to communicate depends on the average local speed $\bar{c}$ with the effectiveness of the interaction also being governed by the mean free path λ.

The first-order nature of the time dependence is tied in to the assumption of a well-defined velocity distribution. To the first approximation all processes are directly determined by existing local velocities and only indirectly by intermolecular forces.

This first-order differential time feature is still preserved in the more sophisticated treatments involving the use of the Boltzmann integrodifferential equation. A continuous distribution function $f(\mathbf{r}, \mathbf{p}, t)$ of both position and momentum is assumed to exist at all times during the process with well-defined initial conditions when appropriate. The existence of such a function presupposes a strong microscopic correlation of velocities from one volume element to another. The second-order time-differential nature of the mechanical problems has been removed at the very start.

If the initial positions and velocities of all the particles were to be specified in a purely random and uncorrelated manner, in general it would not be possible to define a distribution function, let alone speak of its continuity. The function f is not defined as the limit of N/V as V approaches zero, but rather as V approaches some optimum value V_0, which depends on the conditions prevailing in the medium. When V_0 is equal to the total volume, only the average density is obtained; when V_0 approaches zero, fluctuations cause drastic variations from one instant to the next. The situation is somewhat reminiscent of an asymptotic series, whose best value is obtained from the evaluation of a prescribed number of terms after which a divergence occurs.

A distribution function for an arbitrary specification of coordinates and momenta would converge to such an asymptotic limit for no value of V_0, and the Boltzmann method would fail to describe the initial behavior of such a system. In this case one would have to solve the second-order system of equations for the specific array of molecules. This mathematical catastrophe is avoided in that such detailed information about molecular motion is never known; then all problems in rate processes consist in determining the rate of passage of a system from one distribution function to another.

The question of initial conditions in chemical reactions is not a trivial one. It may be supposed that even in the case of the fastest reactions there exists an initial continuous distribution function. Unfortunately, techniques like solving the Boltzmann equation are not too amenable to strongly interacting systems in which, for example, N separate atoms are destined to become nearly $N/2$ molecules with only a few independent atoms in equilibrium.

An important point to remember is that the ordinary rate constant for a reaction is determined in a uniform mixture of reactants. The standard techniques for nonuniform gases are not applicable in their original form. A different type of distribution function is needed to describe the correlation of nearest neighbors. Classically there is no distinction between strongly interacting atoms and diatomic molecules in highly excited vibrational states. The use of a Lennard-Jones potential, $V = (a/R^n) - (b/R^m)$, has been successful in the treatment of transport properties of nonpolar gases.

A correlation function $f(\Delta R, t)$ giving the probability of nearest neighbor separation as a function of time could be determined by methods similar to the Boltzmann equation. Even with a very accurate potential-energy function, the method would fail to describe the reaction $2H \rightarrow H_2$, since the equilibrium probability distribution is governed by quantum effects. At ordinary temperatures for H_2, $\hbar\omega > kT$; and, as outlined earlier, the probability function for the separation of atoms reduces to the ground state wave function of the oscillator, independent of temperature.

This fact illustrates the need for separating motion into classical and nonclassical modes. The hydrogen example would appear to have all the disadvantages of both classical and quantum motion. The problem is at best a difficult one in scattering and capture. From considerations such as these, it will be realized that if all the potential-energy surfaces for the relative coordinates of a molecule were replaced by classical ones without the vibrational quantum levels, not only reaction kinetics but molecular stability and equilibrium statistics would take on a vastly different complexion. It would not be possible to have one hydrocarbon for any appreciable length of time without a thermodynamic equilibrium of all the others. Again, this fact is illustrated by the example of the quantum statistics of the harmonic oscillator.

Since the half-lives of most reactions are considerably longer than collision intervals, it is evident that there is some kind of a bottleneck which allows only a fraction of the collisions to be effective. An important stride forward was made by introducing the concept of the activated complex with the energy barrier. Emphasis was laid on the supposition that there is a quasi equilibrium between reactant molecules and a transient composite species whose half-life is long compared with collision intervals. The methods of equilibrium statistical mechanics were used to find the concentration of such activated complexes provided with certain assumptions about their structure. It was then recognized that even in the process of a reaction, molecules generally tend to retain a large amount of their original character. This led to the proposal that reactions proceed mainly along a primary path, called the reaction coordinate, in which reactants are continuously transformed into products by varying a single relative coordinate along the path of least resistance to such a decomposition.

The time factor was introduced by supposing a distribution of velocities along this reaction path such that the number of states with momenta P in the interval dP and with coordinate Q in the interval dQ is

$$e^{-P^2/2mkT}\frac{dPdQ}{h}.$$

This corresponds to the high quantum number assumption. The existence of a barrier to motion along Q of height $V_0 = P_0^2/2m$ further restricts the possible reacting complexes to those for which $P > P_0$. Integration over all such momenta to give the average rate of crossing the barrier introduces the factor $e^{-V_0/kT}$, so crucial in describing the temperature dependence of rate constants.

It was also recognized that not all complexes with $P > P_0$ will decompose into products. This follows from the wave-mechanical behavior of scattering by repulsive barriers. A transmission coefficient κ is needed to complete the computation of the specific rate constant. In its simplest form the theory produced the relation

$$K = \kappa \frac{F^\ddagger}{F_1F_2\cdots}\frac{kT}{h}e^{-V_0/kT}, \qquad [9.2]$$

where F_1, F_2, etc., are the partition functions of the reactant molecules and $F^\ddagger$ is that of the activated complex.

The ratio of the partition functions comes from the quasi-equilibrium hypothesis; h is a scaling factor for the number of states in a given region of phase space; $kTe^{-V_0/kT}$ is a measure of the average kinetic energy available for crossing a barrier of height V_0; and κ is a wave-mechanical transmission coefficient, which depends on the geometry of the decomposing complex viewed as a scattering phenomenon. In this theory the determination of a rate constant reduces to the determination of three parameters, κ, $F\ddagger$, and V_0.

The quantities $F\ddagger$ and V_0 are determined by the geometry of the activated complex and may be regarded as static properties of quantum-mechanical standing waves. (It should be mentioned that $F\ddagger$ refers to all relative coordinates of the complex except the reaction coordinate, which has been assumed to have the limiting density of states, $1/h$.) Although the specific introduction of time has come through the appropriate average value of P, the transmission coefficient is a dynamical quantity intimately bound up with the details of the decomposition. It is the ratio of the current densities and is, in a sense, a ratio of times.

We now wish to examine more carefully the preceding development. Before expecting miracles from such enormous simplification it is well to attempt an outline of the actual events expected to occur in the course of a reaction. If the simplest case of a binary encounter between two molecules is first considered, it is evident that a whole statistical distribution of collision geometries will prevail. From one point of view in a binary-reaction mixture every molecule and its nearest neighbor of the opposite kind constitute a possible activated complex, since of necessity the complex is not a stable molecule with well-defined dimensions.

The essence of the activated-complex theory seems to lie in the assumption that after a very brief transient state the molecules spend much more time in binary and higher-order encounters than would be predicted by statistics. Rather than being related to the weak interactions between most molecules, the situation is a consequence of the intermediate array of atoms. Even though the complex is unstable it has tied up the atoms in such a way that a high amount of energy is needed to break them apart, except along the so-called reaction coordinate. The theory assumes that all molecules are either existing independently in a region of weak interaction or in one of strong interaction in the form of complexes.

The conventional methods of equilibrium thermodynamics are applicable to molecules in their lowest vibrational states. When there is enough energy in the system for dissociation, the molecules begin to lose their identity. Barriers are broken down to such a degree that new species and new equilibria are introduced.

Alternatives to the above procedure are quite staggering, but there are times when a bare outline of a procedure can be illuminating.

The discussion will be confined to two reacting molecules involved in a one-step, second-order reaction, first order in each, $A + B \rightarrow C + D$. The outline of the program is as follows:

1. Statistical methods are used to obtain the probability of all complexions of the system which have not crossed a barrier. This will give the most probable disposition of the system in the quasi-equilibrium stage of the reaction.
2. For the case of equal numbers of A and B molecules the above statistical analysis will be used to describe the probability distribution of nearest neighbors paired as A and B in such a way that those pairs with the highest probability of collision are favored. This will entail factors depending both on relative velocity and distance.
3. The pairs are considered to be removed to infinite separation and the rate is a weighted average in such a way that those pairs with a large separation will have a small probability, since the probability of two molecules crossing a long distance without intervening collisions is quite low.
4. The crux of the problem is to use time-dependent Schrödinger theory to obtain the average time for such a pair of encountering molecules to produce a suitable distribution of C and D at the average distance of separation that would prevail in mixture of pure C and D.

Needless to say, steps 1–3 are a Herculean task, but there is hope that the powerful methods of statistics such as steepest descent will provide a suitable estimate We wish now to concentrate the bulk of our attention on the temporal aspect given in step 4.

In classical mechanics, a problem is completely defined by a Hamiltonian, $H(Q_i, P_i)$, and initial values of coordinates and conjugate momenta, $Q_i^{(0)}, P_i^{(0)}$. In wave mechanics, instead of a discrete set of initial values one must prescribe an initial wave function giving the initial probability of finding the particle along with an initial current density giving the probability of its motion.

Consider a single degree of freedom with a Hamiltonian, $H(Q, (\hbar/i)(\partial/\partial Q))$. With the usual boundary conditions $\psi(Q) \to 0$ as $Q \to \pm\infty$, there exists a complete set of functions for an attractive potential $\psi_n(Q)$. The most general initial condition is the specification of a complex function $\Theta(Q)$ such that

$$\int_{-\infty}^{\infty} \Theta^*(Q)\Theta(Q)\, dQ = 1.$$

This gives an initial probability density $\rho(Q) = \Theta^*(Q)\Theta(Q)$, with a current density

$$J(Q) = \frac{\hbar}{2mi}\left[\Theta \frac{d\Theta}{dQ} - \Theta \frac{d\Theta^*}{dQ}\right]. \qquad [9.3]$$

It is especially convenient to express $\Theta(Q)$ in the form $\Theta(Q) = \chi(Q)e^{i\theta(Q)}$. Then one obtains

$$\rho(Q) = \chi^2(Q), \qquad [9.4a]$$

$$J(Q) = \chi^2(Q)\,\frac{\hbar}{m}\,\frac{d\theta(Q)}{dQ}. \qquad [9.4b]$$

If $\theta(Q) = P_0Q/\hbar$, the subsequent formulas become particularly simple. From the relation $H = -(\hbar^2/2m)(d^2/dQ^2) + V(Q)$, it follows that

$$E = \int_{-\infty}^{\infty} e^{-iP_0Q/\hbar}\psi_n(Q)He^{iP_0Q/\hbar}\psi_n(Q)\, dQ = E_n + \frac{P_0^2}{2m}. \qquad [9.5]$$

This corresponds to a traveling wave with drift velocity P_0/m having the same probability distribution as one of the stationary states.

The development in time is completely determined by expanding the initial function in terms of the stationary states:

$$\chi(Q)e^{i\theta(Q)} = \sum_n a_n \psi_n(Q), \qquad [9.6]$$

where the a_n are complex numbers with $\sum_n a_n^* a_n = 1$. Even if $\chi(Q) = \psi_n(Q)$, the product $\psi_n(Q)e^{i\theta(Q)}$ is not in general an eigenfunction of H. The time-dependent function

$$\Psi(Q, t) = \sum a_n \psi_n(Q) e^{-iE_n t/\hbar} \qquad [9.7]$$

is the solution to the equation $H\Psi = i\hbar(\partial\Psi/\partial t)$, satisfying the initial condition $\Psi(Q, 0) = \chi(Q)e^{i\theta(Q)}$, provided H is independent of time.

As an incidental diversion, this point of view may be used to convince nonbelievers of the consistency of wave mechanics with classical mechanics in the limit of large mass. If V is taken to be the harmonic potential for the oscillator, the classical and wave-mechanical problems take on the following form:

A. Classical:

Initially $P = P_0$, $Q = Q_0$.

Solution $Q = Q_0 \cos \omega t + \dfrac{P_0}{m\omega} \sin \omega t$.

$$\omega = \frac{1}{2\pi}\sqrt{\frac{k}{m}}.$$

B. Quantum mechanical:

Initially $\quad \theta(Q) = \sqrt{\dfrac{\pi}{8}}\, e^{iP_0 Q} e^{-\gamma(Q-Q_0)^2}.$

Solution $\Psi(Q, t) = \sum_n a_n \psi_n(\sqrt{\beta Q}) e^{-iE_n t/\hbar}$,

where

$$a_n = \int_{-\infty}^{\infty} \psi_n(\sqrt{\beta}\, Q) e^{iP_0 Q} e^{-\gamma(Q-Q_0)^2}\, dQ,$$

$$\beta = \frac{\sqrt{km}}{\hbar},$$

$$E_n = (n + \tfrac{1}{2})\hbar \sqrt{\frac{k}{m}},$$

and

$$\psi_n = \left(\frac{\sqrt{\beta/\pi}}{2^n n!}\right)^{1/2} H_n(\sqrt{\beta}\, Q) e^{-\beta Q^2/2}.$$

When m is large, the Gaussian function $e^{-\beta Q^2/2}$ becomes sharply peaked about the origin. If γ is taken to be $\beta/2$, the expansion for a_n is particularly simple. The summation over n may be replaced by an integral and the resulting probability density $\Psi^*\Psi$ will be seen to represent a traveling wave of the approximate form

$$\kappa(t) e^{-\varepsilon(t,Q)} \left[Q_0 \cos \omega t + \frac{P_0}{m\omega} \sin \omega t\right],$$

where

$$\int_{-\infty}^{\infty} \kappa(t) e^{-\varepsilon(t,Q)}\, dQ = 1$$

and $e^{-\varepsilon(t,Q)}$ is peaked sharply around

$$Q = Q_0 \cos \omega t + \frac{P_0}{m\omega} \sin \omega t.$$

The details of the calculation are space consuming, but the relevance to reaction theory is worth noting. At high temperatures the translational motion of the molecules in a gas may be regarded as nearly classical. The dynamics of nonuniform gases for transport processes as worked out under classical assumptions gives good agreement with experiment in most cases with only small quantum corrections.

We may therefore expect that as two reacting molecules approach, the relative motion will be described by a classical, sharply peaked, traveling wave. When the potential energy starts to rise sharply prior to crossing the reaction barrier, the motion is no longer classical; however, if the molecules are essentially incapable of reaction, they will recoil from one another unchanged. This corresponds to an elastic collision governed by repulsive exchange forces. When the temperature is not too high, such collisions may be regarded as classical with a simple repulsive potential. At higher temperatures one must allow for special quantum effects.

The situation is analogous to elastic and inelastic nuclear scattering. At low energies the scattering is elastic and for larger nucleons, like alpha particles, can be treated classically. At higher energies the specific forces on the nucleus must be considered and the dynamical equations of quantum theory invoked. Thus in a chemical reaction, that part of the collision in which bonds are being broken and re-formed must be regarded as a nonclassical inelastic collision. It seems inconceivable that an accurate comprehensive theory of reaction rates will be forthcoming without specific use of the dynamical quantum equations.

It now remains to outline the path that such an approach may possibly take. A statistical analysis will provide the most probable complexion of initial configurations which have not crossed the barrier. If all possible complexions are included, the calculation reduces to the equilibrium one in which the density matrix is diagonal and independent of time.

We will concentrate on the strong-interaction region where the wave function of the two molecules becomes substantially different from a properly antisymmetrized sum of products of the individual molecular wave functions.

There will exist a complete set of stationary states for any group of atoms which satisfy the eigenvalue equation $H\psi_n = E_n\psi_n$, where

H is the total Hamiltonian of the system. This will not only include all molecules made up of all the atoms but minimum energy conformations of all combinations of the atoms as well. For example, there are two thermodynamically stable species with the formula $C_2H_4Br_2$, 1, 1 dibromoethane and 1, 2 dibromoethane, in addition to equilibrium conformations of such combinations as $Br_2 + C_2H_4$, $HBr + C_2H_4Br$, and $C_2H_2 + 2HBr$. All these systems have the same Hamiltonian, and only the combination of all the eigenfunctions will form a complete set.

When it has been established that a reaction is of a certain order, it means that of all the possible types of collision, only one combination is making a significant contribution to the rate. Generally, complex reactions tend to proceed by a series of binary encounters, each with its own rate constant, rather than by a single or small number of higher-order interactions. In the most precise calculations all possible interactions must be analyzed including the role of intermediate species.

In determining the initial complexions one must speak in terms of the probability of finding a specified number of molecules within a given volume element. This will be related to the thermodynamic theory of fluctuations. In a large volume the fractional change in the number of molecules from one instant to the next is very small. If $\bar{N}$ is the average number of molecules in a small volume, the probability that there are N molecules is

$$\frac{\bar{N}^N e^{-\bar{N}}}{N!}.$$

We have $\bar{N} \sim (V/V_0)N_0$, where N_0 is the total number of particles in the volume V_0. In a gas the volume of the region of strong interaction for a group of molecules is very small compared with the volume per molecule, and the probability of high-order interactions decreases exponentially.

If the analysis stopped here, it would imply that the rate of a second-order reaction was proportional not only to the product of the concentrations but to an exponential factor in the concentrations as well. The key was to recognize that the reaction quickly proceeds from this probability controlled stage to the quantum-mechanical bottleneck where reactants are being tied up in a much smaller volume than predicted by random encounters. It has been assumed that all molecules are tied up in the form of relatively undistorted species or in the form of strongly bound but decomposing complexes. For example, even though the energy of an unstable aggregate of atoms is higher than the sum of the energies of its stable components, on the average it takes still more energy to decompose it in an arbitrary manner. This fact has been behind the success of the quasi-equilibrium theory.

The quantum-mechanical initial states must be specified by an initial configuration of atoms prescribed by a suitable statistical analysis. It is assumed that an electronic wave function for the fixed nuclei can be found. In general this will not even be an estimate to one of the stationary state functions. The normal coordinates for the motion are now given suitably weighted Gaussian probability distributions corresponding to values that would be prescribed by ordinary molecular vibrations. Finally, the initial motion is determined by a statistical array of phase factors leading to a function

$$\Psi^{(0)} = \Phi_{\text{elec}} e^{i\Sigma_{i=1}^{N} P_i^{(0)} Q_i/\hbar} \chi(Q_1) \cdots \chi(Q_N). \qquad [9.8]$$

This is then expanded in terms of the complete set of stationary states for the selected system of atoms:

$$\Psi^{(0)} = \sum a_n \psi_n.$$

And the subsequent development is followed by the exact solution to the problem:

$$\Psi(t) = \sum a_n \psi_n e^{-iE_n t/\hbar}.$$

The specification of these initial states is not completely arbitrary, since they must be consistent with those obtainable from a continuous deformation of reactants without crossing any barriers. The time it takes for decomposition into products will be found by expanding $\Psi(t)$ in terms of a complete and continuous set of functions representing all possible combinations of decomposition in the form of traveling waves, including a return to the original reactants. In the limit, as $t \to \infty$, this will represent a specific one of the infinite set of possible decompositions. The rate of decomposition will then be measured by the rate of growth of the coefficient of the final traveling wave in the general expansion of $\Psi(t)$.

Needless to say, an uncompromising vigorous mathematical attack would quickly disenchant anyone with this formalism. The key to success will be in the anticipation of the final result and the concentration on estimates to wave functions which describe the principal part of the behavior. The devising of some simple even if not completely realistic models will help considerably in determining what road to take.

Modified Equilibrium Reaction-Rate Theory

A cornerstone in the theory of reaction kinetics is the use of the difference in mass between electrons and nuclei as justification for the Born-Oppenheimer approximation, which treats the nuclei as stationary in calculating the potential in which the nuclei move.

Chemiluminescent reactions for which rapid motion along the reaction coordinate leads to jumping from a lower to a closely adjacent potential surface exemplify exceptional behavior. Such cases are, however, successfully treated by time-dependent perturbation theory. Adopting the Born-Oppenheimer approximation, chemistry can then be formulated in terms of variable mass points moving on a potential energy surface in configuration space. Low regions correspond to compounds and systems, and a saddle point separating such basins or valleys is the activated complex of absolute rate theory. Association reactions correspond to descent of the main point from a plateau in such a way as to end up in a valley. Since the motion of a system of n atoms can be treated as a trajectory in $(3n - 5)$-dimensional space for linear molecules and in $(3n - 6)$-dimensional space for nonlinear molecules, mechanical behavior rapidly goes beyond our power to visualize it unless we severely restrict the number of degrees of freedom $3n$ to be considered at any one time. Ordinarily, even in a condensed system, a chemical reaction involves significant changes only in the position of atoms in a restricted number of degrees of freedom so that a potential surface need only be drawn as a function of the distances which change appreciably during reaction. A further simplification arises if we note that certain of the atoms on the periphery of the reaction zone, which change their position comparatively little as the reaction proceeds, can have their effect on the reaction expressed in terms of activity coefficients for the initial and the activated state. In this way a chemical reaction can be usefully described in terms of a potential surface in configuration space involving only a very few well-chosen coordinates.

A wave packet describing a mass point in an initial state designated by an appropriate quantum number will then pass from the initial to the final state by either penetrating or surmounting the barrier, or the wave packet may be reflected. Thus the rate at which any distribution of initial states approaches equilibrium can be treated as a purely quantum-mechanical problem if we are willing to face up to the enormously detailed calculations this procedure involves. Various experimental procedures exist for preparing comparatively simple and reasonably well specified initial states and for observing the final states of the products.

Crossed molecular beams lead to extraordinarily complex scattering problems. By varying the initial conditions and observing the corresponding angles of scattering of products, we get an indication of the forces of interaction in a chemical reaction. Ideally this would show systems in an initial state colliding and passing to various final states. Although much more nearly monoenergetic beams can be prepared than are found in ordinary thermal reactions, the experimental systems are still complicated, and the interpretation, though rewarding, is still far short of the detail which treating each quantum state as a separate entity entails.

An intuitive feeling for the different ways that the energy of reaction can be distributed among the degrees of freedom of the products of reaction can be obtained by studying the best potential surfaces that can be constructed in configuration space. The simplest case for study is a system of three atoms colliding along a line. If energy is plotted vertically while the two distances of the outer atoms from the central atom are plotted as abscissa and ordinate in the horizontal plane, we have the familiar landscape in which two valleys paralleling the two coordinate axes are joined through a saddle point near the origin. By an appropriate choice of scale along the ordinate and a change of the angle between abscissa and ordinate, the motion of the three atoms in collision can be represented by a mass point moving on the potential energy surface. When the three atoms are equal in mass, the angle between ordinate and abscissa must be chosen to be sixty degrees instead of the usual ninety, and the scale of distance along the ordinate should be equal to that along the abscissa. In general, if the atoms be numbered in order, one to three, with the distance r_{23} measured along the abscissa and r_{12} along the ordinate, the angle which r_{12} makes with the vertical should be decreased from ninety degrees by an angle θ, where $\sin\theta = \{(m_1 m_3)/[(m_1 + m_2)(m_2 + m_3)]\}^{1/2}$ and the distances r_{12} should be divided by $\{[m_1(m_2 + m_3)]/[m_3(m_1 + m_2)]\}^{1/2}$ before plotting along the tilted axis.

In the symmetrical case, the reaction trajectory of lowest activation energy passes symmetrically through a simple saddle point so that the trajectory for products duplicates that of reactants. If a shallow basin lies between two saddle points, the lifetime for the resulting metastable state will show up in a molecular beam experiment as scattering over wide angles. Such a shallow basin has no detectable effect on ordinary reaction rates. If for unlike atoms the barrier comes before the midpoint of the two valleys, the products will show high vibrational energy with little translation. On the other hand, if the barrier occurs in the reactant valley, the heat of reaction will appear as translational energy of the products.

When more than three atoms are involved, the interpretation is more complicated but no less interesting. It is still true that if the saddle point occurs in the reactant valley, the energy will be liberated largely into the degrees of freedom of the products and will yield an energy-rich product spectrum. Sufficiently exothermic reactions of this kind give rise to chemiluminescence and even ionization. Stated another way, if, in a reaction, bonds form that release energy before the system breaks into products, the energy will appear as vibrational energy in the products. On the other hand, if the activated complex starts separating into products before the energy of reactant bonds is released, the released energy will be fed into the translational energy of the products.

Symmetry Considerations

The activated complex can be related to reactants and also to products by correlation diagrams using symmetry considerations, in the same way Milliken related the united atom to the two atoms which come together to form the united atom. Thus two states of activated complex with eigenfunctions that are bases of the same irreducible representation will have energy surfaces which do not cross as the activated complex changes toward products or reactants. On the other hand, if the two eigenfunctions are bases of different irreducible representations, their potential energy surfaces may cross. Crossing over from the potential energy surface corresponding to one irreducible representation to a surface corresponding to a different irreducible representation is strictly forbidden if the Hamiltonian is completely invariant under the group operations. In general, complete invariance of the Hamiltonian under the group of operations does not occur, so that crossing over is possible. If this crossing occurs in the neighborhood of the saddle point, the system will have difficulty crossing over to the lowest surface. The necessary electronic reorganization accompanying a crossover will slow the rate of reaction into the crossed-over state.

If two sets of products can be formed from the same set of reactants, those products will be favored that involve the lowest activation energy, unless a change in symmetry near the activated state occurs and so increases the chemical inertia that continuing on the higher energy path occurs. The Woodward-Hoffman set of orbital symmetry rules for organic reactions is one interesting application of such symmetry considerations. Many other applications are possible.

Reaction Rates and Thermodynamics

Van't Hoff pointed out that equilibrium is a dynamic process and that forward and backward reactions balance. According to the principle of detailed balance, this is true for each elementary reaction. If the concentration of activated complexes in molecules per cc per length δ along the reaction coordinate at the saddle point for a reaction is given by $C_\delta^\ddagger$, then the forward rate of reaction R_f is

$$R_f = \tfrac{1}{2}\kappa C_\delta^\ddagger \frac{u}{\delta} \equiv \tfrac{1}{2}\kappa C^\ddagger \frac{(2\pi m^\ddagger kT)^{1/2}\delta}{h} \frac{\sqrt{2kT/\pi m^\ddagger}}{\delta}$$

$$= \kappa \frac{kT}{h} C^\ddagger = \kappa \frac{kT}{h} F^\ddagger \lambda^\ddagger. \qquad [9.9]$$

Here $u = (2kT/\pi m^\ddagger)^{1/2}$ is the average reaction rate along the reaction coordinate, and $C^\ddagger$ is the concentration of activated complexes per quantum state along the reaction coordinate; κ, k, T, h, $m^\ddagger$, $F^\ddagger$, and $\lambda^\ddagger$ are the transmission coefficient, Boltzmann's constant, absolute temperature, Planck's constant, effective mass along the reaction coordinate, partition function of the activated complex, and the absolute activity, respectively. If we have a reaction given by the equations

$$aA + bB \rightarrow C^{\ddagger} \qquad [9.10]$$

$$\lambda_A^a \lambda_B^b = \lambda^{\ddagger}, \qquad [9.11]$$

then

$$R_f = \kappa \frac{kT}{h} F^{\ddagger} \lambda_A^a \lambda_B^b = \kappa \frac{kT}{h} F^{\ddagger} e^{a\mu_A/kT} e^{b\mu_B/kT}$$

$$= \kappa \frac{kT}{h} F^{\ddagger} e^{\{[a(\partial \ln A/\partial n_A)]/kT\}} e^{\{[b(\partial \ln B/\partial n_B)]/kT\}}. \qquad [9.12]$$

Here λ_i and μ_i are the absolute activity and chemical potential, respectively; A is the Helmholtz free energy for the system. If there is a slowly changing quasi equilibrium, it will be a good approximation to take

$$\lambda_A = \frac{C_A}{F_A}; \qquad [9.13]$$

and we have

$$R_f = \kappa \frac{kT}{h} \frac{F^{\ddagger}}{F_A^a F_B^b} C_A^a C_B^b = \kappa \frac{kT}{h} K^{\ddagger} C_A^a C_B^b = \kappa \frac{kT}{h} e^{-\Delta G^{\ddagger}/RT} C_A^a C_B^b$$

$$\equiv k' C_A^a C_B^b. \qquad [9.14]$$

Here $K^{\ddagger}$ is an equilibrium constant for activation, $\Delta G^{\ddagger}$ is the Gibbs free energy of activation, and k' is the specific rate of reaction.

Our use of the forward rate of reaction at equilibrium away from equilibrium is most easily justified by remembering that the backward rate of reaction R_b is without effect on R_f. This is because activated complexes moving in the forward direction decompose independently of other activated complexes. They have no way of knowing whether the activated complexes of the back reaction are suppressed or are proceeding at the equilibrium rate.

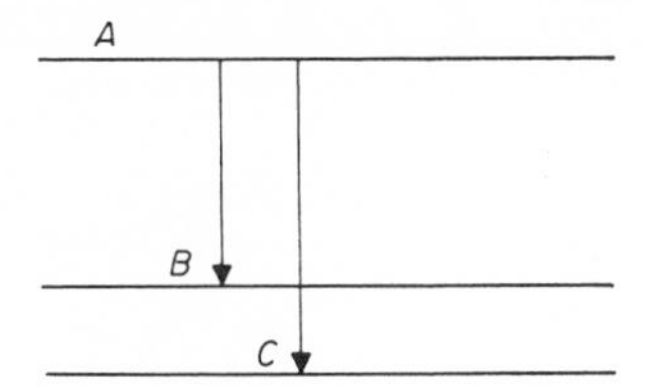

Figure 9.1 Quantum transitions of molecules in quantum state A.

It follows that molecular beam experiments, while interesting and revealing of the course of reaction, are not useful in testing the equilibrium hypothesis of reaction rates.

If one treated all quantum states as species and had usable specific rate constants for passage between all pairs of states, one could avoid all statistical-mechanical considerations in reaction rate theory. Our considerations are reminiscent of Einstein's use of equilibrium conditions to relate spontaneous to induced emission.

This raises the question whether the specific rate of passage between states at equilibrium applies to systems out of equilibrium. This can be tested experimentally. If molecules in a quantum state A pass to both B and C as in Fig. 9.1, one asks the question: Is the ratio of line intensities independent of how state A is excited? In other words, are the relative probabilities of a system in A passing to B or C completely independent of where the molecules in A come from? This is probably only approximately true, so that the calculation of reaction rates involves the consideration of an infinite regression. Fortunately, such experiments as are available indicate that specific rate constants for elementary processes can be used to successfully describe all kinds of simple and complex reactions.

Acknowledgments

We wish to thank the National Science Foundation, the National Institutes of Health, and the U.S. Army Research for support of this research.

Ten

O. Costa de Beauregard

Statistical Irreversibility and Quantized Wave Retardation*

The two related theories that heat is a special form of energy and that the unidirectional transformation of work into heat in isothermal situations is merely one example (like card shuffling) of innumerable probability-increasing transitions must certainly rank among the finest achievements of classical theoretical physics.

But a worrisome question arises: If heat merely represents a maximum of disorder, i.e. of probability, in an ensemble of time-reversible mechanical evolutions, how does it happen that physical systems always evolve in the direction of states of maximum probability? Because of the time symmetry of the postulated equations of motion, should there not be just as many probability-decreasing as probability-increasing evolutions? Stated in this form, the question goes far beyond the domain of statistical mechanics; it concerns the whole area covered by the physical applications of probability theory. One may ask: If, for example, the probability of exchanging two cards in a deck depends only on their positions, why is it that in an overwhelming number of cases shuffling of the deck will produce disorder by destroying order, and not the other way around?

To this question a technical answer was given by Bayes as early as 1763;[1] but it can be safely said that even today the full implications of the one-way prescription for the temporal application of Bayes's principle are not yet understood. Here we have one of the rare instances where common sense is misleading, prattling as it does when referring to obvious observations, the explanation of which is *not* to be found in their mere commonness. So, finally, the time arrow is prescribed in Bayes's principle, just as it was in the principles of Carnot or Clausius (or in Boltzmann's answer to Loschmidt's paradox): *It is simply a principle of physics that, on the*

* Dedicated to the acute analyst of the wave propagation of probability amplitudes.

macroscopic scale, transitions increase the probability, and not vice versa. And as van der Waals pointed out first, the statistical interpretation of Carnot's principle is merely an instance of the temporal application of Bayes's principle.[2]

Let us recall the Bayes formula

$$p_i' = \frac{\bar{\omega}_i p_i}{\sum \bar{\omega}_j p_j},$$

where the p_i denote intrinsic probabilities in a given statistical context, and the $\bar{\omega}_i$ extrinsic or a priori probabilities of the same occurrences. The point is that in physics the extrinsic $\bar{\omega}_i$ must not be used in prediction but in retrodiction; i.e., as Watanabe put it, that *blind prediction is physical, while blind retrodiction is not.*[3,4] Before him, J. W. Gibbs had written: "It should not be forgotten, when our ensembles are chosen to illustrate the probabilities of events in the real world, that while the probabilities of subsequent events may often be determined from those of prior events, it is rarely the case that probabilities of prior events can be determined from those of subsequent events, for we are rarely justified in excluding the consideration of the antecedent probability of the prior events."[5]

It is thus clear that the root of the physical statistical irreversibility is not found in the elementary laws of evolution* but is expressed as a boundary condition for integrating the statistical laws of evolution. As Mehlberg puts it, *physical irreversibility is of a factlike, rather than lawlike, character.*[6]

The physical interpretation of the $\bar{\omega}_i$ in Bayes's formula is now obvious: They represent at best the interaction out of which the statistical system under study has been segregated; and stating that they should be used only in retrodiction is a way of saying that, in

* Today one should be alert to the implications of *PC*, and possibly *T*, violations in some elementary particle decays. To our knowledge, however, no truly convincing argument has been adduced for deriving so general a phenomenon as physical irreversibility from such rare and weak instances as *PC* violations. On the other hand, we are purposely limiting our present discourse to extracting important consequences from physical generalities—a task so brilliantly illustrated by Alfred Landé.

the statistical sense (that is, on a macroscopic scale), physical interactions produce aftereffects, not "before-effects." Hence the classical denomination of Bayes's principle as a *principle of probability of causes*. This important aspect of things has been thoroughly discussed by von Weiszäcker,[7] Reichenbach,[8] Grünbaum,[9] this writer,[10,11] and others listed in the references.[12–16]

In physics there exists another very wide class of temporally dissymmetric phenomena—a class where the theoretical formulation of this dissymmetry is also given in the form of a one-way road sign. This is the class of wave-propagation phenomena, where, in the macroscopical sense, the whole semiclass of advanced solutions is excluded by decree, in the form of a *boundary condition*. In the realm of classical wave propagation, nothing more can be said. There is no explicit hint that some kind of statistics might be involved in wave propagation, and thus no possible proof that a physical connection might exist between the principles of probability increase and of wave retardation.

There are some implicit hints, however. When a stone falls in a pond, or when a meteorite is slowed down in the atmosphere, the fact is that retarded waves (surface, elastic, ballistic, or radiation waves) are mediating the physical process of probability increase. Consider, for instance, a more specific experiment. If, between times t_1 and t_2, a piston in a cylinder containing a gas in equilibrium is moved, Maxwell's velocity law will be disturbed momentarily. The point is that the disturbance begins at time t_1 and does not end at time t_2: this is the anti-Loschmidt, the Carnot, or the Bayes principle. Also, the disturbance is emitted at time t_1 as a retarded wave in the gas, not absorbed at time t_2 as an advanced wave in the gas. So it really seems as if, physically at least, some kind of connection might exist between the two principles of probability increase and wave retardation. That such a connection might be upheld has been denied by Popper.[17] It has, however, been expressed by many physicists, including the present writer.

With the advent of quantized waves things became less academic and more operational. One may recall, in this context, the controversy between Ritz and Einstein,[18] in which Ritz believed that the principle of retarded waves was a presupposition when deducing the Carnot-Clausius law, while Einstein maintained that, on the contrary, it was the probability increase that was implied in wave retardation. May it not be that wave mechanics was just the missing link preventing Einstein and Ritz from recognizing that they were saying the same thing in reciprocal forms? Had they known that particle scattering, in the sense of statistical mechanics, is also wave scattering in the sense of wave mechanics, then it should have been clear to both of them that the quantization of light waves was only half the truth they needed in order to tell the whole story.

Also, from Planck's definition of the entropy of a light beam (in which the photon concept is implied), it follows that the scattering of light entails an entropy increase.[19] Here again (in the special case of Bose statistics) the connection between the two principles, retarded waves and entropy* or probability increase, is obvious. Should wave con-fusion rather than dif-fusion be postulated, then the entropy would decrease.

With the introduction of such concepts as phonons and rotons, the principle of wave quantization may well be taken as universal in wave theory; the postulate of a general connection between the two concepts of wave retardation and probability increase thus seems to be quite sound. It is not my purpose, however, to pretend to deal with so large a program. I merely want to show in two definite instances precisely what I have in mind.

Consider a plane grating which, when receiving a plane monochromatic wave (the wave planes being parallel to the lines on the grating), reemits g outgoing plane waves whose intensities, for simplicity, we assume to be equal. Each of the g outgoing waves may be excited by any of the g ingoing waves of one and the same family, to which, of course, belongs the one considered first.

* When using a logarithmic basis larger than unity.

In terms of quantized waves, the grating induces *transitions* of particles between two sets of orthogonal states (each of them uniquely correlated with the other), all transition amplitudes being assumed equal.

We now suppose, again for simplicity, that the corpuscles impinge on the grating at mean time intervals long enough for them to be individually recognizable.* The number of ways a given distribution of occupation numbers n_i of the g states can be obtained is then

$$P(n_i) = \frac{n!}{\Pi n_i!}\,.$$

It is easily verified that $P(n_i)$ decreases if one transfers a particle from one state to another that is more fully occupied, so that the most probable distribution is the one with all n_i equal, with possible differences of ± 1.

Let us compare two experimental arrangements:

1. All incident particles belong to the same incoming wave; they are emitted from the same collimator or source. Then, a blind statistical prediction assigns equal occupation numbers to the g outgoing waves. The principle of wave retardation yields the same result: equal intensities of outgoing waves in phase-coherent scattering.
2. All outgoing particles that are detected are on the same outgoing wave; they are received in the same telescope or sink. A blind statistical retrodiction would yield the unphysical conclusion that the incident particles are distributed equally among the g possible incoming waves; but this procedure is the one forbidden by Bayes's principle. Similarly, postulating the existence of advanced waves would yield the unphysical conclusion of phase-coherent con-fusion instead of dif-fusion.

The thought experiment just described shows that exactly parallel conclusions are drawn from Bayes's principle and from

* That is, they must not be in the same quantum cell or wave train.

the principle of retarded waves. When speaking of quantized waves in this context, we are in fact dealing with two different wordings of one and the same principle.

The preceding example is, of course, just one of many others yielding a similar conclusion. So it may be interesting to state the general conclusion in an abstract fashion. This is easily achieved by merely rewording von Neumann's well-known theory of irreversibility in the quantum-mechanical measuring procedure.[20]

Let us recall the essential points in von Neumann's demonstration. We write

$$S \equiv \sum \bar{\omega}_i \Psi_i$$

for the density matrix in its diagonal representation, where the Ψ_i denote the projectors corresponding to the orthogonal states ψ_i, and the $\bar{\omega}_i$ are a set of statistical weights such that

$$0 \leq \bar{\omega}_i \leq 1, \qquad \sum \bar{\omega}_i = 1.$$

Then it follows, from the general quantum rules, that the statistical weight of some eigenstate ψ_i' not belonging to the ψ_i set is, after a measurement of the corresponding physical magnitude, given by

$$\bar{\omega}_i' = \sum_j c_{ij}^* c_{ij} \bar{\omega}_j,$$

wherein the c_{ij} are the matrix elements interchanging the ψ_i and ψ_i' representations. Denoting by p the largest $\bar{\omega}_i$ and using the identity

$$\sum_j c_{ij}^* c_{ij} = 1,$$

one finds

$$\bar{\omega}_i' \leq p,$$

which expresses the leveling-out tendency inherent in physical statistics.

Clearly, the procedure used in this demonstration is blind statistical prediction. The point is that it is also a retarded-wave method, because the measuring procedure (ideally supposed to occur at some time t_0) is taken as a source of retarded waves and not as a sink of advanced waves.

Finally, we may summarize the general lesson of both our *Gedankenexperiment* and von Neumann's reasoning in the following concise statement. In quantum theory, use of retarded waves is synonymous with statistical prediction, which is physical; while use of advanced waves is synonymous with statistical retrodiction, which, when used "blindly," is unphysical. So, in the theory of quantized waves, the Bayes principle, asserting that blind retrodiction is forbidden, and the principle by which advanced waves do not exist on the macroscopic level, are indeed one and the same principle in different wordings. For further discussion see the references.[21–25]

We now concentrate on the topical features of the time symmetry inherent in the quantum transition; the main one clearly is that the *individual* quantum transition has no preference whatever for retarded over advanced waves. This we illustrate by using a characteristic example, and one very puzzling to common sense, the Einstein-Podolsky-Rosen paradox.[26]

At the 1927 Solvay Council, Einstein raised the following objection against the statistical interpretation of quantum mechanics.[27] A matter wave, supposed for simplicity to carry just one particle, is diffracted by a small hole O, and subsequently absorbed by a semi-spherical photographic plate. As soon as the particle is received on the plate at point A at time t, one is certain that it is not received at any point $B \neq A$ of the plate. The essential idea is that in a *Gedankenexperiment* the radius of the plate is arbitrarily large, so that the logical inference made by an observer (A) operating around the space-time point (A, t) may concern very large space-time intervals, and moreover, very large spacelike intervals. Clearly, the latter kind of logical inference is neither prediction nor retrodiction; it is *telediction*. And, as Renninger pointed out in 1963,[28] this inference can just as well be of a negative form: If, knowing the time interval during which the particle passes through the hole O and its velocity range, the observer (A) operating around A at the right time t does *not* detect it, then he is sure that the particle hits the plate at some other point $B \neq A$.

A slight modification of this *Gedankenexperiment* will emphasize the completely symmetric roles of two distant observers (A) and (B). If the initial event is the separation, by a semitransparent mirror, of a very weak light beam carrying just one photon per time interval Δt, and the observation consists in detecting the presence or absence of the photons in the two outgoing beams at two very distant places (A) and (B), then the "teledictions" from (A) to (B) and from (B) to (A) are evidently completely symmetric.

While such a state of affairs admittedly presents no mystery whatsoever in a deterministic world picture (something has to be here or there, so if it is here, it is not there), things are not so simple in a probabilistic world picture. And so much the less that, first of all, equal care must be taken of the contingency inherent in probabilism and the objectivity inherent in space-time Minkowskian geometry.* That this somewhat paradoxical synthesis is indeed operational is very well attested in, say, the Tomonaga-Schwinger-Feynman electrodynamics.†

Does the solution to our conceptual difficulties consist merely in the recognition that Minkowski's space-time has no less reality than Euclid's three-dimensional space? The acceptance of true probabilism in a static three-dimensional context does not raise, at first sight, unanswerable questions, as is obvious from the old problems involving balls in boxes or decks of cards. So can we simply say that our particles are distributed on orthogonal waves in space-time,

* The Einstein-Podolsky-Rosen argument *stricto sensu* is formulated in nonrelativistic quantum mechanics with, of course, no upper limit for the velocity of signals. The *Gedankenexperimente* we have chosen easily allow for a relativistic discourse, which naturally is more physical.

† As early as 1929, Hugo Bergmann insisted that quantal probabilism and relativistic objectivity have to be reconciled. His argumentation is mainly philosophical and indeed reflects a very keen insight.

just as balls in boxes are in space; and that the observers simply do not know on which waves the particles are before their proper time has carried them in space-time into the here-now domain where the experiment takes place? It should be remembered that the experiment, together with its result, is a macroscopic event, and as such it is depicted in space-time once and for all according to the postulates of relativity theory.

This philosophy—though yet incomplete, as we shall see—has an important implication. In the objective space-time of relativity theory, information transfers (that is, logical connections pertaining to all the facets of a single statistical occurrence) are geometrical in their expression and have nothing to do with the macroscopic time arrow. But is not *this* precisely the lesson of the Einstein paradox and the like? Do not the very formulas of the calculation show that, in our problem, the information is telegraphed either from (A) to (B) or from (B) to (A), first from present to past, and then from past to future, along a Feynman zigzag having its apex in the space-time domain where "the die is cast"? Does not the Einstein argument prove, in an operationally irrefutable way, that when quantal individual occurrences are at stake, one *can* indeed telegraph into the past, so as finally to produce a space-like "telediction"? Thus, Einstein's former prohibition of telegraphing into the past must be a macroscopic, not a microscopic, prohibition. And this, of course, brings us back to the main concern of our paper: It is, in fact, the last disguise of the Loschmidt kind of paradox.

The preceding discourse is prevented from concluding the story by just one point inherent in the quantum-mechanical measuring process. A quantum measurement is a procedure which is not merely passive; it has an active facet also. If, in the preceding example, neither the (A) nor (B) observers are operating, then the two partial beams remain phase-coherent and may be made to interfere anew. But if, on the other hand, the occupation number is measured at either (A) or (B), then the phase coherence is destroyed and the beams will no more be able to interfere. This, incidentally, shows that a quantal occurrence is very far indeed from a classical, deterministic one and may justify our taking great pains in trying to understand it. This also shows, as Bohr puts it, that the experimental arrangement as a whole contributes in producing the answer to the question it is asking.[29]

So it must be conceded that the *information* gained in a quantum measurement is not merely a *gain in knowledge*; it is also in some sense *producing what comes to be known*. This sounds very Aristotelian indeed, because, for Aristotle, *information* was both *gain in knowledge* and *organizing power*. For further discussion see the references listed.[30–33]

Finally, we return to the vexing question of why blind statistical prediction is physical, while blind statistical retrodiction is not.

Lewis writes that "gain in entropy always means loss of information, and nothing else. It is a subjective concept." The point I wish to make is that if *information* merely meant *knowledge*, this would be too crude a statement. Consider, for instance, the dissipation of the sun's energy through retarded electromagnetic waves. This occurs with or without the presence of scientists, engineers, or ordinary men on earth. Nothing is changed by them unless someone uses his understanding of the phenomenon to recapture part of the dissipative flow by, for example, constructing a barrage or extracting fuel. This is the process of converting *information* into *coarse-grained negentropy*—a process basically similar to the workings of Maxwell's demon, as discussed by Szilard, Demers, Brillouin,[34] and others.[35,36]

Two somewhat symmetrical transitions must therefore be considered. The first is the *learning transition*, which consists in extracting information I from a preexisting coarse-grained negentropy N_1 according to the scheme $N_1 \rightarrow I$; the irreversibility principle here asserts that $N_1 \geq I$. The second is the *acting transition*, which consists in converting one's knowledge or information of the situation into a subsequent coarse-grained negentropy N_2 according to the scheme $I \rightarrow N_2$; the irreversibility principle here states that $I \geq N_2$.

Now it must be kept in mind that the irreversibility principle is of a factlike, rather than lawlike, character. We have just said that gaining knowledge is a procedure associated with increasing entropy, i.e. also with retarded actions; while producing order, i.e. decreasing entropy, should be associated with advanced actions. It thus turns out that *cognizance* and *will* should be the two kinds of awareness associated respectively with the progression and the regression of fluctuations. Thus, to say that retarded actions are statistically outweighing advanced ones amounts to holding that acquiring knowledge is for us far easier than producing order.

But is not this precisely what is implied in the very smallness of Boltzmann's constant k when expressed in practical units? The fundamental interpretation of Boltzmann's constant is that $k \ln 2$ is the equivalence coefficient between a negentropy expressed in thermodynamic units and an information expressed in binary units, according to the formula

$$\Delta N = k \, \Delta I \ln 2.$$

To say that $k \ln 2$ is very small when expressed in practical units amounts to saying that learning is very cheap in negentropy terms, while producing order costs a lot in information terms. Transition to the limit $k \rightarrow 0$ would imply that knowledge is costless and action impossible—a theory well known under the name "epiphenomenal consciousness."

In conclusion, we feel that, in the factlike, rather than lawlike, exclusion of entropy-decreasing or advanced-wave evolutions, some quite fundamental and yet unsuspected problems might well be involved. The positron, after all, which is de facto much rarer than the electron, fills a mathematically de jure open possibility, and its experimental discovery has started quite a bit of new physics.

We also feel that the sui generis probability laws of quantum theory, with their wave propagation of probability amplitudes and the time symmetry inherent in them (as exemplified by the Einstein kind of paradox), imply some consequences for information transfer in space-time that have not yet been fully explored. But why should not a scientist speculate?

References

1.
T. Bayes, "Essay towards Solving a Problem in the Doctrine of Chances," Phil. Trans. Roy. Soc. London (1763).
2.
J. D. van der Waals, Physik. Z. **12**, 547 (1911).
3.
S. Watanabe, Phys. Rev. **97**, 26 (1955).
4.
S. Watanabe, "Le concept de Temps dans le Principe d'Onsager," in *Transport Processes in Statistical Mechanics*, ed. by I. Prigogine (Interscience, 1958), p. 285.
5.
J. W. Gibbs, *Elementary Principles in Statistical Mechanics*, first published in 1902 (Dover, 1960), p. 150.
6.
H. Mehlberg, "Physical Laws and Time's Arrow," in *Current Issues in the Philosophy of Science*, ed. by H. Feigl and G. Maxwell (Holt, Rinehart & Winston, 1961), p. 105.
7.
C. G. von Weiszäcker, Ann. Physik **36**, 275 (1939).
8.
H. Reichenbach, *The Direction of Time* (University of California Press, 1956).
9.
A. Grünbaum, *Philosophical Problems of Space and Time* (Knopf, 1963), pt. 2.
10.
O. Costa de Beauregard, Rev. Synthèse No. **5–6**, 7 (1957).
11.
O. Costa de Beauregard, "Irreversibility Problems," in *Proceedings of the 1964 International Congress for Logic, Methodology, and Philosophy of Science*, Jerusalem, 26 Aug.–2 Sept. 1964. (North-Holland, 1965).
12.
E. N. Adams, Phys. Rev. **120**, 675 (1960).
13.
W. Büchel, *Philosophische Probleme der Physik* (Herder, 1965), chap. 2.
14.
P. and T. Ehrenfest, *The Conceptual Foundations of the Statistical Approach in Mechanics*, trans. of article in *Encykl. Math. Wiss.* (1912) by M. J. Moravcsik (Cornell University Press, 1959).
15.
Ya. P. Terletskii, J. Phys. Radium **21**, 681 (1960).
16.
M. M. Yanase, Ann. Japan Assoc. Phil. Sci. **1**, 131 (1957).

17.
K. R. Popper, Nature **177**, 538 (1958); **178**, 382 (1958); **179**, 1296 (1958); **181**, 402 (1958).
18.
W. Ritz and A. Einstein, Physik. Z. **10**, 323 (1909).
19.
M. Planck, *Theory of Heat* (Macmillan, 1932), pts. 2 and 3.
20.
J. von Neumann, *Mathematical Foundations of Quantum Mechanics* (Princeton University Press, 1955).
21.
O. Costa de Beauregard, Cahiers Phys. **12**, 317 (1958).
22.
J. A. McLennan, Phys. Fluids **3**, 493 (1960).
23.
O. Penrose and I. C. Percival, Proc. Phys. Soc. (London) **79**, 605 (1962).
24.
S. Watanabe, Phys. Rev. **84**, 1008 (1951); **97**, 40 (1955); **97**, 179 (1955).
25.
T. Y. Wu and D. Rivier, Helv. Phys. Acta **34**, 661 (1961).
26.
A. Einstein, B. Podolsky, and N. Rosen, Phys. Rev. **47**, 777 (1935).
27.
A. Einstein, in *Rapports du 5ème Conseil Solvay, Paris, 1928*, p. 253.
28.
M. Renninger, Physik. Z. **136**, 251 (1963).
29.
N. Bohr, Phys. Rev. **48**, 696 (1935).
30.
J. S. Bell, Physics **1**, 195 (1964).
31.
W. Büchel, Physik. Bl. **4**, 162 (1967).
32.
O. Costa de Beauregard, Dialectica **19**, 280 (1965).
33.
W. Furry, Phys. Rev. **49**, 393 (1936).
34.
L. Brillouin, *Science and Information Theory* (Academic Press, 1956).
35.
O. Costa de Beauregard, Ann. N.Y. Acad. Sci. **138**, 407 (1967).
36.
D. Gabor, M.I.T. Lectures (1951). Referred to in Ref. 34, p. 168.

Eleven **Fritz Bopp**

The Internal Symmetries of Elementary Particles Resulting from the Geometric Structure of Lattice Space

1. Formulation of the Problem

The investigations here presented will show that geometric structures can underlie the internal symmetries of elementary particles, as in the analogous case of Lorentz invariance. I shall examine the internal symmetries that derive from a simple cubic lattice with superstructure. As the group of the internal symmetries of the kinetic operator (*Bewegungsoperator*), one obtains $SU(4)$, which contains $SU(3)$ as a subgroup. There exists no interaction operator of an equally high symmetry. The group of maximal attainable symmetry of the interaction operator reads $SO(4) \times (1) \times (1)$. Within this group, the one-parameter groups of the phase transformations and of the Touschek transformations are constructed.

To begin with, we are dealing with purely mathematical assertions, which at present cannot be confronted with experience, as that would require the solutions to many unsolved dynamic problems. However, our considerations open up new correlations of such harmony that even if the lattice chosen here should prove to be inadequate, that per se regrettable fact would not be without meaning for the understanding of the internal symmetries of elementary particles.

We start here, as elsewhere, with a finite lattice space[1] which, for the purpose of clarifying its structure, we imagine to be embedded in a Euclidean space. This assumption has at present merely methodical significance. If at each point there exists only one pair of creation and annihilation operators of the Fermi type, then the Hilbert space associated with the lattice has a finite dimensionality. Hence calculations in this Hilbert space are unproblematical. Only when numerical results are involved, i.e. for masses, effective cross sections, and the like, do we make the transition to the limit, which is carried out in such a fashion that the lattice becomes arbitrarily fine and has any extension required. The lattice's sole function, so to say, is that it enables us to define this transition to the limit.

Yet it may, just because of the connection exhibited here, have a more than methodical significance. Since it will be shown that the specific characteristics of the lattice are reflected in the symmetries of the elementary particles, it is just these characteristics that matter. Therefore, a definite structure will be impressed upon the continuum through the transition to the limit. The continuum enriched with (or spanned by) the lattice thus leads to a theory with a more comprehensive structure than one in Euclidean space; this is already expressed in the fact that there exists a length distinguished by the nature of the limiting transition. The preceding obviously suggests the question as to whether space *in fact* does or does not possess lattice structure. No less a thinker than Riemann [2] already investigated the problem. Naturally one cannot test it directly by experimental means.

The dominating significance of the continuum representation (pictorial or conceptual) is closely related to the basic conception of classical mechanics according to which all physical occurrences should be attributed to the motions of mass points—not conceivable as anything but continuous—which themselves always remain the same. The compulsion thereby emerging toward a continuum representation no longer obtains in quantum mechanics, because of the existence of quantum jumps. For this reason, the question about the structure of space is today an open one. We must keep the possibility in mind that it will someday be established that space is more like a lattice than a Euclidean continuum. Here, however, we regard the lattice merely as a methodical aid.

2. Lattice Space Embedded in the Euclidean Continuum

We base our considerations on the lattice space introduced in Ref. 1 and, moreover, start with a simple cubic lattice whose arbitrarily chosen

$$\text{lattice constant} = a. \qquad [11.1]$$

The lattice vectors are

$$\mathbf{n} = (n_1, n_2, n_3). \qquad [11.2]$$

The finiteness of the lattice enters through our identification of the lattice points modulo Z:

$$\mathbf{n} + Z\mathbf{n}' = \mathbf{n} \ (\text{mod } Z). \qquad [11.3]$$

This lattice, we suppose, is not yet the space lattice; it forms only the scaffolding from which we proceed in the construction of the space lattice. Let the true space lattice be spanned by the following points:

$$\text{space points} = \mathbf{n} + \mathbf{a}_\mu, \qquad \mu = 1, 2, \ldots, 8; \qquad [11.4]$$

the eight vectors $\mathbf{a}_\mu$ are defined by

$$\mathbf{a}_\mu = (\pm\tfrac{1}{4}, \pm\tfrac{1}{4}, \pm\tfrac{1}{4}), \qquad \mu = 1, \ldots, 8. \qquad [11.5]$$

The arbitrary assignment of the indices 1 to 8 to the various combinations of plus and minus signs is carried out as in Fig. 11.1. (The present numbering is chosen differently from that in Ref. 1, as it leads to matrices that one can more easily survey.)

If we consider the lattice points by themselves, they are seen to form a simple cubic lattice with lattice constant $\frac{1}{2}a$. The grouping of eight points in each case to form a lattice cell is done for reasons

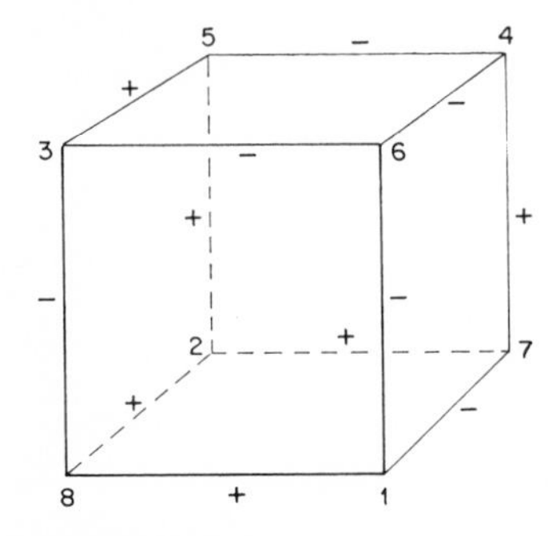

Figure 11.1 Numbering of lattice points of a cell and the appropriate algebraic transition signs.

of dynamics, to which we shall return later. Here we anticipate what remains to be said formally. Plus or minus signs are given to all edges of the lattice cell in Fig. 11.1; we select them in such a way that they are interchanged by reflection at the center of the cell. In addition, at least one of the three axes should remain the same. Accordingly there are only two groupings which do not go over into one another through the symmetry operations of the cubic lattice, and physically not even these prove to be essentially different. Thus we hit upon the choice shown in our diagram.

This picture repeats itself from cell to cell. Between the cells the signs can be chosen in such a manner that they alternate along each of the main lattice lines. In the next section we shall see how the kinetic operator is determined by these signs, and that in the limit $Z \to \infty$ one arrives at the operator of the Dirac equation for particles with zero mass.

First we have to fix the transition to the limit.[3] We proceed from a definite but still unknown length L, with which we define the lattice constant a and the lattice length A as follows:

$$a = LZ^{-1/2} \qquad \text{and} \qquad A = LZ^{1/2}. \qquad [11.6]$$

To this assumption there corresponds a lattice constant b and a lattice length B (or momentum) in momentum space:

$$b = (2\pi\hbar/L)Z^{-1/2} \qquad \text{and} \qquad B = (2\pi\hbar/L)Z^{1/2}. \qquad [11.7]$$

The limiting transition is just chosen in such a manner that both lattice constants a and b approach zero in the same manner—which, in view of the canonical symmetry, one can hardly imagine any other way. As to the size L, we make no assumptions, but rather choose L, $\hbar$, and c as units.

3. The Dirac Equation in Lattice Space

Let the lattice space contain particles of a type we call "primordial fermions" or "urfermions" (*Urfermionen*). Accordingly, there exists at every point of space a pair of creation and annihilation operators

$$\psi_\mu^\dagger(\mathbf{n}),\ \psi_\mu(\mathbf{n}), \qquad \mu = 1, 2, \ldots, 8,$$

which satisfy the Jordan-Wigner commutation relations. In the Schrödinger picture, which we use here, they are independent of time. It is always possible to find representations in which the Schrödinger operator has the form[4]

$$H = (1) + (\psi^\dagger\psi) + (\psi^\dagger\psi^\dagger\psi\psi) + \cdots. \qquad [11.8]$$

Herein, the parentheses signify homogeneous forms of the kind given within the parentheses. In these representations the number of particles is constant, and only such representations will be allowed in the sequel.

If the universe as a whole is governed by a single Schrödinger operator, i.e. by a world-function, as Mie[5] first proposed—at that time still within the framework of classical physics and with the use of classical electromagnetic fields—then the Schrödinger operator furnishes a series of independent world-types, which are differentiated from one another by the number of urfermions. On the assumption of *C*-symmetry, one type distinguishes itself by a particularly high symmetry, namely that symmetry for which exactly half of the lattice points are occupied. We confine our attention to this world-model with $4Z^3$ urfermions and arrive thus, with a single assumption, at the Dirac-sea conception, which is always possible in a finite-dimensional Hilbert space and as such should not be rejected.

One may consequently make use of the language of solid state physics. The whole universe can be likened to a Fermi liquid of urfermions whose ground state is the Dirac sea. The real world would consist of locally bounded excitation levels, which are comparable with the quasi particles in the Fermi liquid. Elementary particles accordingly have to be quasi particles in the world-wide Dirac sea; borrowing a metaphor from Schrödinger, one could speak of foam tops in the sea. Whether or not these actually exist depends upon the choice of the Schrödinger operator.

This operator is for practical purposes determined uniquely by the lattice. Let us demonstrate this property first for the kinetic operator. To this end, it is expedient to number temporarily the lattice points, determined by $\mathbf{n}$ and μ, in arbitrary order with the index i running from 1 to $8Z^3$, so that one can write $\psi_i^\dagger$ and ψ_i respectively for the creation and annihilation operators. Allowing transitions only between neighbors, we assume that the transition amplitudes are constant, on account of the homogeneity of space, and choose the transition signs in accordance with Fig. 11.1; then the kinetic operator is given by

$$H_B = \sum_{i,k} (\pm)\psi_i^\dagger \psi_k, \qquad [11.9]$$

wherein the summation extends over neighboring pairs.

If we return to the operators $\psi_\mu^\dagger(\mathbf{n})$ and $\psi_\mu(\mathbf{n})$ and proceed from there to the Fourier components $\tilde{\psi}_\mu^\dagger(\mathbf{h})$ and $\tilde{\psi}_\mu(\mathbf{h})$, where $\mathbf{h}$ is a reciprocal lattice vector, the simple result

$$H_B = \sum_h \tilde{\psi}^\dagger(\mathbf{h})(\rho_1 \boldsymbol{\sigma} \cdot \mathbf{k} + \rho_2 \boldsymbol{\tau} \cdot \boldsymbol{\kappa})\psi(\mathbf{h}), \qquad [11.10]$$

comparable with the Dirac operator, is then obtained. Herein $\tilde{\psi}^\dagger(\mathbf{h})$ and $\tilde{\psi}(\mathbf{h})$ are eight-component spinors. The symbols $\boldsymbol{\rho}, \boldsymbol{\sigma}, \boldsymbol{\tau}$ represent three commuting 8×8 matrices, which behave like Pauli matrices and exhibit the symmetry properties

$$\boldsymbol{\rho}^T = -\rho_2 \boldsymbol{\rho} \rho_2, \qquad \boldsymbol{\sigma}^T = -\boldsymbol{\sigma}, \qquad \boldsymbol{\tau}^T = -\boldsymbol{\tau}. \qquad [11.11]$$

Figure 11.1 determines the matrices $\boldsymbol{\rho}, \boldsymbol{\sigma}, \boldsymbol{\tau}$, which are

$$\boldsymbol{\rho} = \boldsymbol{\rho}'; \qquad \boldsymbol{\sigma} = (\tau_2'\sigma_1', \sigma_2', \tau_2'\sigma_3'); \qquad \boldsymbol{\tau} = (-\sigma_2'\tau_1', \sigma_2'\tau_3', \tau_2'). \qquad [11.12]$$

The matrices $\boldsymbol{\rho}', \boldsymbol{\sigma}', \boldsymbol{\tau}'$ appearing here are the following Kronecker products in ordinary Pauli matrices $\boldsymbol{\sigma}_P$:

$$\boldsymbol{\rho}' = \boldsymbol{\sigma}_P \times 1 \times 1, \qquad \boldsymbol{\sigma}' = 1 \times \boldsymbol{\sigma}_P \times 1, \qquad \boldsymbol{\tau}' = 1 \times 1 \times \boldsymbol{\sigma}_P. \qquad [11.13]$$

The lattice structure is contained in the vectorlike quantities $\mathbf{k}$ and $\boldsymbol{\kappa}$:

$$k_i = Z^{1/2} \sin \frac{2\pi h_i}{Z}, \qquad \kappa_i = Z^{1/2}\left(1 - \cos \frac{2\pi h_i}{Z}\right). \qquad [11.14]$$

Here $\mathbf{k}$ is the momentum and, for finite momenta and sufficiently large Z, $\boldsymbol{\kappa}$ is vanishingly small. If the term in $\boldsymbol{\kappa}$ is omitted in [11.10], we are left with the eight-component Dirac equation.

The sole difference is that $\mathbf{k}$, according to Eq. [11.14], can take on only a finite number of values. These lie in the interval

$$-Z^{1/2} \leq k_i \leq Z^{1/2}, \tag{11.15}$$

and their separation is

$$\frac{dk_i}{dh} = 2\pi Z^{-1/2} \cos \frac{2\pi h_i}{Z}. \tag{11.16}$$

If we start from values of $\mathbf{k}$ and $\boldsymbol{\kappa}$ that are compatible with Eq. [11.14], then generally Lorentz transformations will lead to values which no longer satisfy this equation. We can, however, always choose Z so large that within immeasurably small intervals there are still as many lattice points as desired. In consequence, provided that Z is sufficiently large, deviations from Lorentz invariance can never be discerned. In this sense, the lattice space theory is "quasi-Lorentz-invariant" for large enough Z.

4. Internal Symmetries of the Kinetic Operator

We consider the operator in Eq. [11.10] without the $\boldsymbol{\kappa}$-term:

$$H_B = \sum_h \tilde{\psi}^\dagger(\mathbf{h})\rho_1\boldsymbol{\sigma}\cdot\mathbf{k}\psi(\mathbf{h}). \tag{11.17}$$

Internal symmetries are characterized by the transformations of the operators $\tilde{\psi}^\dagger$ and $\tilde{\psi}$ alone which leave the Schrödinger operator invariant. Taking account of transformations of the Pauli-Gürsey type,[6] we start from the *ansatz*

$$\psi(\mathbf{n}) \to \psi'(\mathbf{n}) = A\psi(\mathbf{n}) + B\psi^{\dagger\!\dagger}(\mathbf{n}), \tag{11.18}$$

with $\psi^{\dagger\!\dagger} = \psi^{\dagger T}$. The 8×8 matrices A and B are, for reasons of homogeneity, independent of $\mathbf{n}$, and the Fourier components are, correspondingly,

$$\tilde{\psi}(\mathbf{h}) \to \tilde{\psi}'(\mathbf{h}) = A\tilde{\psi}(\mathbf{h}) + B\tilde{\psi}^{\dagger\!\dagger}(-\mathbf{h}). \tag{11.19}$$

The matrices A and B should be so defined that transformed operators too satisfy the Jordan-Wigner commutation relations and that the Schrödinger operator remains invariant.

From the invariance of the commutation relations, the conditions

$$AA^{\dagger} + BB^{\dagger} = 1 \qquad \text{and} \qquad A^{T}B + BA^{T} = 0 \tag{11.20}$$

follow. If we supplement Eq. [11.19] by its Hermitian conjugate equation, in which we moreover replace $\mathbf{h}$ by $-\mathbf{h}$,

$$\psi'^{\dagger\dagger}(-\mathbf{h}) = B^{*}\tilde{\psi}(\mathbf{h}) + A^{*}\tilde{\psi}^{\dagger\dagger}(-\mathbf{h}), \tag{11.21}$$

then we can combine both transformations in a sixteen-component equation; with the use of the notations

$$\hat{\psi}(\mathbf{h}) = \begin{pmatrix} \tilde{\psi}(\mathbf{h}) \\ \tilde{\psi}^{\dagger\dagger}(\mathbf{h}) \end{pmatrix}, \qquad \hat{U} = \begin{pmatrix} A & B \\ B^{*} & A^{*} \end{pmatrix}, \tag{11.22}$$

this equation reads

$$\hat{\psi}'(\mathbf{h}) = \hat{U}\hat{\psi}(\mathbf{h}). \tag{11.23}$$

Herein $\hat{U}$ is unitary, according to Eq. [11.20], i.e.

$$\hat{U}\hat{U}^{\dagger} = \hat{U}^{\dagger}\hat{U} = 1, \tag{11.24}$$

but, on account of [11.22], in a special way. If one sets

$$\boldsymbol{\lambda} = \boldsymbol{\sigma}_{P} \times 1 \times 1 \times 1, \tag{11.25}$$

then

$$\lambda_1 \hat{U} \lambda_1 = \hat{U}^{*} \tag{11.26}$$

must hold. Obviously, these special transformations also form a group.

We are mainly interested in the continuous part of this group, and hence go out from the infinitesimal transformations

$$\hat{U} = 1 + i\delta\hat{\xi} \qquad \text{and} \qquad \delta\hat{\xi}^{\dagger} = \delta\hat{\xi}. \tag{11.27}$$

If we put

$$\delta\hat{\xi} = \xi_0 + \boldsymbol{\lambda}\cdot\boldsymbol{\xi} = \begin{pmatrix} \xi_0 + \xi_3 & \xi_1 - i\xi_2 \\ \xi_1 + i\xi_2 & \xi_0 - \xi_3 \end{pmatrix}, \qquad [11.28]$$

it follows from the Hermitian property

$$\xi_\mu^\dagger = \xi_\mu, \qquad \mu = 0, 1, 2, 3, \qquad [11.29]$$

and from [11.26] that

$$\xi_0^* = \xi_0^T = -\xi_0, \qquad \xi_1^* = \xi_1^T = -\xi_1,$$

$$\xi_2^* = \xi_2^T = -\xi_2, \qquad \xi_3^* = \xi_3^T = \xi_3. \qquad [11.30]$$

By [11.30], ξ_3 is symmetric and all other ξ_μ are antisymmetric.

To test the commutability of the operator [11.17] with H, we write it as

$$H_B = \tfrac{1}{2}\sum_{\mathbf{h}} \{\tilde{\psi}^\dagger(\mathbf{h})\rho_1\boldsymbol{\sigma}\cdot\mathbf{k}\tilde{\psi}(\mathbf{h}) - \tilde{\psi}^T(-\mathbf{h})\rho_1\boldsymbol{\sigma}\cdot\mathbf{k}\tilde{\psi}^{\dagger\dagger}(-\mathbf{h})\}.$$

This can be rewritten as

$$H_B = \tfrac{1}{2}\sum_{\mathbf{h}} \hat{\psi}^\dagger(\mathbf{h})\lambda_3\rho_1\boldsymbol{\sigma}\cdot\mathbf{k}\hat{\psi}(\mathbf{h}).$$

It follows that H_B is invariant under the transformation $\hat{U}$ if

$$\hat{U}^\dagger\lambda_3\rho_1\boldsymbol{\sigma}\cdot\mathbf{k}\hat{U} = \lambda_3\rho_1\boldsymbol{\sigma}\cdot\mathbf{k} \qquad [11.31]$$

and

$$[\delta\hat{\xi}, \lambda_3\rho_1\boldsymbol{\sigma}\cdot\mathbf{k}] = 0. \qquad [11.32]$$

Since the foregoing must hold for all $\mathbf{k}$, the ξ_μ can not depend on $\boldsymbol{\sigma}$. Thus, from the last equation,

$$[\delta\hat{\xi}, \lambda_3\rho_1] = 0. \qquad [11.33]$$

Of course, any $\boldsymbol{\tau}$ is possible; $\boldsymbol{\lambda}$ and $\boldsymbol{\rho}$ can, however, appear only in the combinations

$$(\boldsymbol{\tau}, \rho_1\boldsymbol{\tau}; \lambda_1\rho_2, \lambda_1\rho_3\boldsymbol{\tau}; \lambda_2\rho_2, \lambda_2\rho_3\boldsymbol{\tau}; \lambda_3, \lambda_3\rho_1). \qquad [11.34]$$

These are sixteen linearly independent infinitesimal transformations, presenting a sixteen-parameter group. By [11.33], $\lambda_3\rho_1$ commutes with all the remaining ones. This matrix produces the one-parameter group of the Touschek transformations.

The algebra constructed from the products of 1, $\boldsymbol{\rho}$, $\boldsymbol{\sigma}$, $\boldsymbol{\tau}$ can be represented by 8×8 matrices. If we regard $\lambda_3\rho_1$ as a diagonal matrix whose eigenvalues are arranged according to magnitude, then the matrices commuting with it have two 4×4 "boxes" along the main diagonals. Because of relations [11.29] and [11.30], the content of the lower box is determined by that of the upper. We are therefore left with a 4×4 matrix whose elements correspond in a one-one manner to the coefficients of the linear form [11.34]. Accordingly, the group defined by the infinitesimal transformations [11.34] is equivalent to the group of the unitary transformations in four-dimensional space. The Touschek group is still to be supplemented by the group $SU(4)$, which contains $SU(3)$ as a subgroup.

We shall not investigate here the discrete transformations. There exists P-, C-, and T-invariance even under inclusion of the $\boldsymbol{\kappa}$-terms.

5. The Interaction Operator

On restricting oneself to local two-particle interactions, the interaction operator (*Wechselwirkungsoperator*) can be written as

$$H_W = \sum_{\mathbf{n},i,k} c_{ik}(\psi^\dagger(\mathbf{n})\beta_i\psi^{\dagger\dagger}(\mathbf{n}))(\psi^T(\mathbf{n})\beta_k\psi(\mathbf{n})). \qquad [11.35]$$

This agrees formally with the usual local postulates and means that we call everything "local" that takes place inside a lattice cell. This procedure is, in view of our space model, not completely consistent. The errors are, however, of the order of magnitude $Z^{-1/2}$, so that we may adhere to operators of the type [11.35].

We have written the operator in normal order. By means of a Fierz transformation,[7] one can return at any time to the more customary current representation. But the latter is, because of the familiar adopted identities, not uniquely determined—in contrast to operator [11.35]. The difference originates in the fact that from the outset only skew symmetric matrices occur in [11.35]:

$$\beta_i^T = -\beta_i. \qquad [11.36]$$

In what follows, it is convenient to use the notation

$$W = \sum_{i,k} c_{ik}\beta_i \circ \beta_k \qquad [11.37]$$

for the interaction matrix appearing in [11.35].

We want to determine this matrix by making use of Heisenberg's requirement[8] that H_W should have the same symmetry as H_B. Here again one arrives at a unique result. To be sure, the maximum attainable symmetry is less than that of H_B; the symmetries stemming from $SU(4)$ and $SU(3)$ are breached by the interaction operator.

From the Lorentz invariance and the $\boldsymbol{\tau}$-invariance, required by [11.34], five different interaction matrices result. Except for an arbitrary factor, they read:

$$B_1 = \boldsymbol{\tau} \circ \boldsymbol{\tau},$$

$$B_2 = -\rho_3\boldsymbol{\sigma} \circ \rho_3\boldsymbol{\sigma} + \rho_2 \circ \rho_2,$$

$$B_3 = \boldsymbol{\sigma} \circ \boldsymbol{\sigma} - \rho_1\boldsymbol{\sigma} \circ \rho_1\boldsymbol{\sigma},$$

$$B^4 = \rho_2\boldsymbol{\sigma}\boldsymbol{\tau} \circ \rho_2\boldsymbol{\sigma}\boldsymbol{\tau} - \rho_3\boldsymbol{\tau} \circ \rho_3\boldsymbol{\tau},$$

$$B^5 = -\rho_1\boldsymbol{\tau} \circ \rho_1\boldsymbol{\tau}; \qquad [11.38]$$

herein $\boldsymbol{\sigma}$ and $\boldsymbol{\tau}$ are multiplied scalarly with themselves. These are the familiar invariants of the four-component theory, of which none vanishes, because the symmetrical ones among them are made skew symmetric by insertion of a factor $\boldsymbol{\tau}$.

Let us now investigate the influence of the remaining transformations which belong to matrix A in Eq. [11.18]. Among these are those infinitesimal transformations from Eq. [11.34] that are independent of λ_1 and λ_2; they include especially, besides the $\boldsymbol{\tau}$-transformations already considered, the phase transformation defined by λ_3, under which H_W is trivially invariant. The earlier noted matrix $\lambda_3\rho_1$ yields the Touschek transformation

$$\delta\psi = i\rho_1\psi, \qquad \delta\psi^{\#} = -i\rho_1\psi^{\#}.$$

Accordingly, we have

$$\delta(\psi^T A\psi) = (\psi^T \delta A\psi) = i\psi^T\{\rho_1, A\}\psi,$$
$$\delta(\psi^\dagger A\psi^{\#}) = (\psi^\dagger \delta A\psi^{\#}) = -i\psi^\dagger\{\rho_1, A\}\psi^{\#},$$

which gives

$$\delta B_1 = 2i(\boldsymbol{\tau} \circ \rho_1\boldsymbol{\tau} - \rho_1\boldsymbol{\tau} \circ \boldsymbol{\tau}) \neq 0,$$
$$\delta B_5 = 2i(\boldsymbol{\tau} \circ \rho_1\boldsymbol{\tau} - \rho_1\boldsymbol{\tau} \circ \boldsymbol{\tau}) \neq 0. \qquad [11.39]$$

Both operators formed with B_1 and B_5 are therefore no longer separately invariant. Their sum remains, however, unchanged:

$$\delta(B_1 + B_5) = 0. \qquad [11.40]$$

Then B_3 drops out as noninvariant:

$$\delta B_3 = 2i(\boldsymbol{\sigma} \circ \rho_1\boldsymbol{\sigma} - \rho_1\boldsymbol{\sigma} \circ \boldsymbol{\sigma} - \rho_1\boldsymbol{\sigma} \circ \boldsymbol{\sigma} + \boldsymbol{\sigma} \circ \rho_1\boldsymbol{\sigma}) \neq 0.$$

On the other hand, because of $\rho_1\rho_i + \rho_i\rho_1 = 0$, the matrices B_2 and B_4 are Touschek-invariant for $i = 1, 2$:

$$\delta B_2 = 0 \qquad \text{and} \qquad \delta B_4 = 0.$$

The keystone of our argument is furnished by the transformations $\rho_1\boldsymbol{\tau}$. By virtue of the rotation symmetry in $\boldsymbol{\tau}$-space, it suffices to consider one of these, say $\rho_1\tau_1$. The result is

$$\delta\psi = i\rho_1\tau_1\psi, \qquad \delta\psi^{\#} = i\rho_1\tau_1\psi^{\#},$$
$$\delta\psi^T = -i\psi^T\rho_1\tau_1, \quad \delta\psi^\dagger = -i\psi^\dagger\rho_1\tau_1.$$

From this, we get

$$\delta(\psi^T A\psi) = -i\psi^T[\rho_1\tau_1, A]\psi,$$
$$\delta(\psi^\dagger A\psi^{\#}) = -i\psi^\dagger[\rho_1\tau_1, A]\psi^{\#},$$

or, abbreviated,

$$\delta A = -i[\rho_1\tau_1, A]. \qquad [11.41]$$

By this we obtain for $B_1 + B_5$,

$$\delta(B_1 + B_5) = 2(\tau_2 \circ \rho_1\tau_3 - \tau_3 \circ \rho_1\tau_2 + \rho_1\tau_3 \circ \tau_2 - \rho_1\tau_2 \circ \tau_3 - \rho_1\tau_2 \circ \tau_3 + \rho_1\tau_3 \circ \tau_2 - \tau_3 \circ \rho_1\tau_2 + \tau_2 \circ \rho_1\tau_3) \neq 0,$$

so that this matrix too drops out. Similarly, we find for B_2 and B_4:

$$\delta B_2 = 2(\rho_3\boldsymbol{\sigma} \circ \rho_2\tau_1\boldsymbol{\sigma} + \rho_2 \circ \rho_3\tau_1 + \rho_2\tau_1\boldsymbol{\sigma} \circ \rho_3\boldsymbol{\sigma} + \rho_3\tau_1 \circ \rho_2),$$

$$\delta B_4 = 2(\rho_2\tau_1\boldsymbol{\sigma} \circ \rho_3\boldsymbol{\sigma} + \rho_3\tau_1 \circ \rho_2 + \rho_3\boldsymbol{\sigma} \circ \rho_2\tau_1\boldsymbol{\sigma} + \rho_2 \circ \rho_3\tau_1).$$

Both matrices are likewise eliminated individually. Their difference is, however, invariant:

$$\delta(B_2 - B_4) = 0. \qquad [11.42]$$

Accordingly, a single interaction operator remains, viz.

$$H_W = W \sum_{\mathbf{n}} \{-(\psi^\dagger\rho_3\boldsymbol{\sigma}\psi^{\dagger\dagger})(\psi^T\rho_3\boldsymbol{\sigma}\psi) + (\psi^\dagger\rho_2\psi^{\dagger\dagger})(\psi^T\rho_2\psi) - (\psi^\dagger\rho_2\boldsymbol{\sigma}\boldsymbol{\tau}\psi^{\dagger\dagger})(\psi^T\rho_2\boldsymbol{\sigma}\boldsymbol{\tau}\psi) + (\psi^\dagger\rho_3\boldsymbol{\tau}\psi^{\dagger\dagger})(\psi^T\rho_3\boldsymbol{\tau}\psi)\}. \qquad [11.43]$$

Since in the case of infinitesimal transformations of the Pauli-Gürsey type, expressions of the form $(\psi^\dagger\psi^\dagger)(\psi\psi)$ become $(\psi^\dagger\psi^\dagger)(\psi^\dagger\psi)$ or $(\psi^\dagger\psi)(\psi\psi)$, invariance can only then obtain if the individual factors are invariant. Let us consider first of all the transformations formed with $\lambda_1\rho_2$ and $\lambda_2\rho_2$ from Eq. [11.34]:

$$\delta'\psi = i\rho_2\psi^{\dagger\dagger} \qquad \text{and} \qquad \delta''\psi = \rho_2\psi^{\dagger\dagger}.$$

They give

$$\delta'(\psi^T A\psi) = i(\psi^T A\rho_2\psi^{\dagger\dagger} - \psi^\dagger\rho_2 A\psi) \neq 0,$$

$$\delta''(\psi^T A\psi) = (\psi^T A\rho_2\psi^{\dagger\dagger} - \psi^\dagger\rho_2 A\psi) \neq 0.$$

Thus $\lambda_1\rho_2$ and $\lambda_2\rho_2$ are eliminated as symmetry operations. The same holds for the remaining transformations of the Pauli-Gürsey type, $\lambda_1\rho_3\boldsymbol{\tau}$ and $\lambda_2\rho_3\boldsymbol{\tau}$, since one can produce them from $\lambda_1\rho_2$

and $\lambda_2\rho_2$ through commutation with $\rho_1\boldsymbol{\tau}$. Consequently, $B = 0$; in [11.34] only the symmetry operations

$$\tfrac{1}{2}(1 \pm \rho_1)(\boldsymbol{\tau}, \lambda_3) \tag{11.44}$$

remain. To the two signs there correspond two rotation groups in $R(3)$ and two one-parameter groups. As a result, we have the group of internal symmetries

$$SO(3) \times SO(3) \times (1) \times (1) = SO(4) \times (1) \times (1) \tag{11.45}$$

in the case of interaction. This is an eight-parameter subgroup of $SU(4)$, but not of $SU(3)$.

The interaction operator is also P-, C-, and T-invariant. Only the C-invariance presents a new aspect. The C-conjugation in our case is

$$\psi \to \rho_3\psi^{\dagger\dagger}. \tag{11.46}$$

From the interaction operator it thus follows that

$$H_W^C = W \sum_{\mathbf{n}} \{-(\psi^T\rho_3\boldsymbol{\sigma}\psi)(\psi^\dagger\rho_3\boldsymbol{\sigma}\psi^{\dagger\dagger}) + (\psi^T\rho_2\psi)(\psi^\dagger\rho_2\psi^{\dagger\dagger}) - \cdots\},$$

that is, in normal representation,

$$H_W^C = H_W + 32W \sum \psi^\dagger\psi - 128WZ^3. \tag{11.47}$$

Obviously, the Schrödinger operator by itself is not C-invariant. However, the urfermion number is constant. The worlds with different numbers of urfermions are completely separated. According to what we assume, we exist in the world containing $4Z^3$ urfermions, i.e.

$$\sum \psi^\dagger\psi = 4Z^3. \tag{11.48}$$

In this world, the last two terms of Eq. [11.47] just cancel one another. Thus, in the actual physical world,

$$H_W^C = H_W, \tag{11.49}$$

so that the C-invariance is assured. It is extraordinarily remarkable how everything precisely comes to a head in a single instance.

6. Summary

1. Space is a finite, simple cubic lattice with a superstructure that is specified by alternating transition signs.
2. At each lattice point there is only one pair of creation and annihilation operators.
3. The kinetic operator is determined by the requirements that only transitions to nearest neighbors occur and that the transition amplitudes are constant except for an alternating sign. In the limit $Z \to \infty$ we obtain the operator of the eight-component Dirac equation.
4. In addition to Lorentz invariance there is internal symmetry. The group of symmetry operations that leaves H_B invariant is isomorphic with $U(4) = SU(4) + (1)$.
5. The interaction operator H_W, when restricted to local two-particle interactions, is uniquely determined by the internal symmetries. However, the complete symmetry of H_B is not attainable. The symmetry group of H_W is $SO(4) \times (1) \times (1)$.
6. P-, C-, and T-invariance exist even when we include the otherwise neglected τ-terms. The C-invariance is tied to the number $4Z^3$ of urfermions.

References

1.
F. Bopp, Z. Physik **205**, 103 (1967).
2.
G. F. Riemann, *Ges. Math. Werke* (Leipzig, 1892), 2d ed., no. XIII, p. 272; with comments by H. Weyl, "Über die Hypothesen welche der Geometrie zugrunde liegen" (Springer, 1920), 2d ed.
3.
F. Bopp, Sitzber. Bayer. Akad. Wiss.. Math.-Naturw. Kl., 1958, p. 220.
4.
F. Bopp, Z. Physik **200**, 142 (1967).
5.
G. Mie, Ann. Physik **37**, 511 (1912); **37**, 1 (1912); **40**, 204 (1912).

6.
W. Pauli, Nuovo Cimento **6**, 204 (1957); F. Gürsey, Nuovo Cimento **7**, 411 (1958).
7.
W. Heisenberg, *Einführung in die Feldtheorie der Elementarteilchen* (Hirzel, 1967), §3-1.
8.
F. L. Bauer, Sitzber. Bayer. Akad. Wiss., Math.-Naturw. Kl., 1952, p. 111.

Twelve **Jean-Pierre Vigier**

Hidden Parameter Theory and Possible Unification of External and Internal Motions of Elementary Particles within $SO(6, 1)$ without Symmetry Breaking

Most of the past examinations of the physical meaning of quantum theory have dealt with (a) the physical significance of the wave field and (b) the complete (or incomplete) character of quantum theory. In the case (b), the discussions have centered mainly on the theoretical possibility of introducing new dispersionless variables (the famous hidden parameters), without making any specific proposal that could be tested experimentally. In the present paper, written in honor of Alfred Landé, whose contribution to this discussion has been so outstanding, we shall attempt a new move in this direction and try to connect the hidden parameters with the new quantum numbers introduced in order to classify elementary particles. As a first step, we shall start with pure group-theoretical considerations, leaving the physical interpretation open for further discussion.

One of the unsolved problems of elementary particle theory is the unification[1,2] of an external group of motion E (describing the particle's behavior in external space-time) and an internal symmetry group I [giving the internal characteristics of $SU(3)$, for example] within a global dynamical group $G(E, I)$ in such a way that the mass splitting within the $SU(3)$ multiplets appears without symmetry breaking.

It has been known for some time that the Poincaré group P cannot be utilized in E for such a unification,[3,4] since the mass spectrum reduces then to a single point. In the present article we shall replace P by $SO(4, 1)$, which does not exhibit this property[5] and is physically indistinguishable from P in the Wigner-Inonu (WI) limit. The group $SO(4, 1)$ has been used by Dirac to describe the electron's behavior in de Sitter space. Its "momentum operators" $P_\mu = \alpha M_{\mu 5}$ ($\alpha = R^{-1}$, R = radius of the de Sitter universe) do not commute. For $SO(4, 1)$ the set of state-labeling operators is thus smaller than for P. There remain only two commuting elements in an Abelian Cartan subalgebra.

The quadratic Casimer operator contains both $P_\mu P^\mu$ and $J^2 - N^2$. In the WI contraction, some unitary representations of $SO(4, 1)$ contract to unitary representations of P. In this process, the corresponding nonunitary finite-dimensional irreducible representations of $SO(4, 1)$ and $SO(3, 1)$ (out of which the unitary representations have been built in Wigner's sense) contract into each other.

We also require that I contains subgroups corresponding to the known physical forces, namely $SU(3)$ (strong interactions), $SU(2) \times SU(2)$ (weak interactions), and $U(1)$ (electromagnetic interactions); since it now appears that weak decays can be accounted for by $SU(2) \times SU(2)$ with electromagnetic perturbations.[6, 7]

For physical reasons, we further impose the following requirements.

a. The complete set of state-labeling operators of the global dynamical group must contain the operators J^2, J_3, T^2, T_3, Y, and μ^2, that is, the total angular momentum, its third component, isospin, its third component, hypercharge, and the squared mass μ^2 with WI limit $P_\mu P^\mu$. Note the absence of the P_1 and P_2 components in this set. The substitution of $SO(4, 1)$ for P thus implies a new quantum measurement theory that gives meaning to the measurement of "nearly commuting operators" in the WI sense.

b. The operators of E should commute with the state-labeling I operators (such as T^2, T_3, Y), because external motion in the absence of interactions should not transform the particle's (internal) nature.

c. There should exist a link between E and I within G, such that μ^2 (which belongs to E's enveloping algebra) splits with respect to internal quantum numbers.

To satisfy all these requirements, we propose here to take $E = SO(4, 1) \times U(1) = g \times U(1)$ [$U(1)$ corresponding to an external electromagnetic field[8]] and $G = SO(6, 1) \times U(1)$; with the essential requirement that the "external" $SO(4, 1)$ (denoted by g) and a certain subgroup $g' = SO(4, 1) \subset SO(6, 1)$ are left and right translations of the same $SO(4, 1)$, that is, are right and left translations of $SO(4, 1) \subset SO(6, 1)$, the $SO(6, 1)$ of $G = U(1) \times SO(6, 1)$.

We shall now show that the main advantage of this assumption is that the quadratic Casimer operator C_1 of g and of g' are identical; so that C_1 does not commute with the $SU(3)$ generators of $SU(3) \subset SU(4) \simeq SO(6) \subset SO(6, 1)$, despite the fact that the g generators commute with all the g' generators and, more generally, with all generators of I. This satisfies requirements b and c. Clearly, $SO(6, 1)$ contains $SU(3) \subset SU(4) \simeq SO(6)$ and $SU(2) \times SU(2) \subset SU(4)$ as subgroups, the baryon number group being associated with the $U(1)$ in $G = U(1) \times SO(6, 1)$.

The exact meaning of the utilization of both types of "translations" of $SO(4, 1)$ can be physically interpreted. If we attach in the case of $SO(3)$, for example, a three-pode "internal" frame L to the particle (particle frame) and a three-pode frame L' to the "external" observer (observer frame), then the relative angular momenta of L and L' can be projected either on L [yielding J_k operators with $(J_i, J_j) = -iJ_k$] or on L' [yielding J'_k with $(J'_i, J'_j) = -iJ'_k$], J_k and J'_k representing the right and left translations of the same $SO(3)$ group defined on the group three-parameter manifold, such as the Euler angles $\omega(\varphi, \theta, \psi)$. Of course, we get $C_1 = J_kJ^k = J'_kJ'^k$ and $(J_i, J'_j) = 0$. The wave fields ($s = \sin\theta/2$, $c = \cos\theta/2$)

$$Y_l^{mm'}(\varphi, \theta, \psi) = s^{-m+m'}c^{-m-m'} \frac{d^{l-m}}{d^{l-m}s^{l-m}} \times [s^{2(l-m')}c^{2(l+m)}] \exp i(m\varphi + m'\psi)$$

depend on two sets of "magnetic" quantum numbers m and m' but on the same l, corresponding to the common Casimer operator C_1. If we fix l and m (or l and m'), we obtain representations $D'(l)$ or

$D(l)$ of $SO(3)$. The application of $R(J_k)$ or $R'(J'_k)$ to the four-component spinor of $l = \frac{1}{2}$, for example, preserves the m or m', respectively. The application of the external J'_k does not change the internal magnetic numbers m which characterize the particle's nature. Physically, the frames L and L' in E_3 can be replaced, according to Yukawa's multilocal model, by two sets of three points on S_2. This means that we parametrize the isometry transformations on S_2 not with two angles θ, φ $[ds^2 = R^2\,(d\theta^2 + \cos^2\theta\, d\varphi^2)]$ but with three angles

$$\theta,\ \varphi,\ \psi\ [ds^2 = R^2\,(d\theta^2 + \cos^2\theta\, d\varphi^2 + \sin^2\theta\, d\psi^2)],$$

which characterize the application of S'_2 on S_2, S'_2 being a moving S_2 surface with respect to a fixed S_2.

This procedure can immediately be generalized to any $SO(p, 1)$ group.

Before we calculate the mass formula, we must first classify the particles in the finite irreducible nonunitary representations of G. Since $G = U(1) \times SO(6, 1)$, we first distinguish particles with different values of the baryon number eigenvalue of this $U(1)$, namely $B = 0, \pm 1, \pm 2$. As $SO(6, 1)$ is a group of rank three, we can—following step by step a method first proposed by Lovelace[9] for $SO(7)$—employ Dynkin's procedure[10] to construct these representations within the group's weight diagram, denoting the three simple roots by $\boldsymbol{\alpha}$, $\boldsymbol{\beta}$, $\boldsymbol{\gamma}$; these are vectors that define an oblique coordinate system in a three-dimensional space. Introducing then the covariant coordinates Y (hypercharge), T_3 or Q (charge), and L (leptocharge), so that in this space $\mathbf{x} = Y\boldsymbol{\alpha} + Q\boldsymbol{\beta} + L\boldsymbol{\gamma}$, we know that every irreducible representation of $SO(6, 1)$ is uniquely defined by the assignment of nonnegative integer values to the three circles of the Dynkin diagram 0—0═0, which can be regarded as the combination of the $SO(4, 1)$ diagram 0═0 with the $SU(3)$ diagram 0—0. This implies that we can split the $SO(6, 1)$ finite nonunitary representation in the weight space into simultaneous decompositions of $SO(4, 1)$ and $SU(3)$. For fixed $\pm Y$ values we get conjugate $SU(3)$ representations in parallel $\boldsymbol{\alpha\beta}$ planes; and for fixed $\pm L$ values we get identical $SO(4, 1)$ representations in $\boldsymbol{\beta\gamma}$ planes. Their intersection (along $\boldsymbol{\beta}$) defines the isobaric spin $T = SU(2)$ group.

If we relate $\boldsymbol{\alpha}$, $\boldsymbol{\beta}$, $\boldsymbol{\gamma}$ to Y, T_3, L, the usual values Y', and T_3' are obtained by displacement of the origin into the $\boldsymbol{\alpha\beta}$ $SU(3)$ planes through the relations

$$T_3' = T_3 \qquad \text{and} \qquad \frac{Y}{2} = \frac{Y'}{2} + \frac{L}{3}.$$

If we further introduce the charge

$$Q = T_3 + \frac{Y}{2} + \frac{L}{3},$$

one sees that $Q' = Q - 2L/3$, so that

$$Q' = T_3' + \frac{Y'}{2}.$$

Of course, we can change the meaning and orientation of the $\boldsymbol{\beta}$ axis and work with the values Y, Q, L (T_3 being defined by the relation for Q above) to find in a simpler manner the Weyl-Cartan representation of $SO(6, 1)$.

On setting $\gamma^2 = 1$, the scalar products of the three simple roots in weight space are $(\mu, \nu = \alpha, \beta, \gamma)$

$$\begin{array}{cc} & \begin{array}{ccc} \alpha & \beta & \gamma \end{array} \\ g^{\mu\nu} = (\boldsymbol{\mu}, \boldsymbol{\nu}) = & \begin{pmatrix} 2 & -1 & 0 \\ -1 & 2 & -1 \\ 0 & -1 & 1 \end{pmatrix}. \end{array}$$

We can then write the eighteen roots in the general form $\mathbf{i} = \sum_\mu a_\mu' \boldsymbol{\mu}$ and verify that they are $\pm\boldsymbol{\alpha}$, $\pm\boldsymbol{\beta}$, $\pm\boldsymbol{\gamma}$, $\pm(\boldsymbol{\alpha} + \boldsymbol{\beta})$, $\pm(\boldsymbol{\beta} + \boldsymbol{\gamma})$, $\pm(\boldsymbol{\alpha} + \boldsymbol{\beta} + \boldsymbol{\gamma})$, $\pm(\boldsymbol{\beta} + 2\boldsymbol{\gamma})$, $\pm(\boldsymbol{\alpha} + \boldsymbol{\beta} + 2\boldsymbol{\gamma})$, $\pm(\boldsymbol{\alpha} + 2\boldsymbol{\beta} + 2\boldsymbol{\gamma})$.

Denoting by H_μ the operators with eigenvalues Y, Q, L, we introduce the operators $H^\mu = g^{\mu\nu}H_\nu$ and obtain the usual set of relations

$$(H_\mu, H_\nu) = 0, \qquad (H_\mu, E^i) = a_\mu^i E^i,$$
$$(E^i, E^{-i}) = \sum_\mu a_\mu^i H^\mu, \qquad (E^i, E^j) = N^{ij} E^{i+j},$$

where $N^{i+j} \neq 0$ only if $i + j$ is a root.

Introducing furthermore the "generalized $SO(6, 1)$ free spin" $S_\alpha = \frac{1}{16}\, \varepsilon_{\alpha\beta\gamma\delta\sigma\rho\varepsilon} M^{\beta\gamma} M^{\delta\sigma} M^{\rho\varepsilon}$, we classify the particles in the "odd" and "even" spin representations, writing after Dynkin:

$$(Y, Q, L) = \begin{pmatrix} 1 & 1 & 1 \\ 1 & 2 & 2 \\ 1 & 2 & 3 \end{pmatrix} \begin{pmatrix} l \\ m \\ \frac{n}{2} \end{pmatrix}.$$

We can describe the structure of the "odd" representation as in Table 12.1, where in the rows the values of the $SU(3)$ and $SO(4, 1)$

Table 12.1. Structure of the lower "odd" (Fermi) representations of $SO(6, 1)$

(l, m, n)	Dimension	Decomposition under $SU(3)$					Decomposition under $SO(4, 1)$			
		$L = \frac{1}{2}$	$L = \frac{3}{2}$	$L = \frac{5}{2}$	$L = \frac{7}{2}$	$L = \frac{9}{2}$	$Y = \frac{1}{2}$	$Y = \frac{3}{2}$	$Y = \frac{5}{2}$	$Y = \frac{7}{2}$
(0, 0, 1)	8	3	1				4			
		6*								
(1, 0, 1)	48	3	8	3*			16	4		
		3	1				4			
		15	8	6	3		20	16		
(0, 1, 1)	112	6*	8	3*			16			
		3	1				4			
		3								
		15	10	6	3	1	20	20		
(0, 0, 3)	112′	6*	8	3*			16			
		3	1							

Table 12.2. Structure of the lower "even" (Bose) representations of $SO(6, 1)$

		Decomposition under $SU(3)$				Decomposition under $SO(4, 1)$		
(l, m, n)	Dimension	$L = 0$	$L = 1$	$L = 3$	$L = 5$	$Y = 0$	$Y = 1$	$Y = 2$
(0, 0, 0)	1	1				1		
(1, 0, 0)	7	1	3*			5	1	
(0, 1, 0)	21	8	3*	3		10	5	
		1				1		
(2, 0, 0)	27	8	3*	6*		14	5	1
		1				1		
(0, 0, 2)	35	8	6	3	1	10	10	
		1	3*			5		

Table 12.3. Multiplication table

1	7	8	21
7	27 + 21 + 1	48 + 8	105 + 35 + 7
8	48 + 8	35 + 21 + 7 + 1	
21	105 + 35 + 7	112 + 48 + 8	

representations are indicated by the corresponding values of L and Y. Only $L > 0$ and $Y > 0$ are represented, since one gets for $-L$ the conjugate $SU(3)$ multiplets and for $-Y$ the identical $SO(4, 1)$ multiplets. We note (as a straightforward consequence of Dynkin's construction) that "odd" $SO(6, 1)$ representations contain only "odd" $SO(4, 1)$ and $SO(3)$ representations. Thus Table 12.1 represents only fermions. The product of "odd" representations yields "even" $SO(6, 1)$, "even" $SO(4, 1)$, and $SO(3)$ representations, as shown in Table 12.2, which represents bosons. Its multiplication results are given in Table 12.3.

We see that the fundamental representation $\{8\}$ (out of which any representation can be constructed as a product) splits into two

$SU(3)$ triplets ($L = \pm\frac{1}{2}$) and two $SU(3)$ singlets ($L = \pm\frac{3}{2}$), all with $SO(4, 1)$ spin $\frac{1}{2}$ (see Table 12.3). Any particles of the scheme can be obtained by "fusion" of such particles according to de Broglie's original idea.

Of course, $SO(4, 1)$ spin $S_\alpha = \frac{1}{8}\varepsilon_{\alpha\beta\gamma\delta\sigma}M^{\beta\gamma}M^{\delta\sigma}$, which goes in the WI limit over into the Poincaré spin $S_\alpha = \frac{1}{2}\varepsilon_{\alpha\beta\gamma\delta}P^\beta M^{\gamma\delta}$, is not constant within all $SU(3)$ multiplets. Thus only certain sets of Tables 12.1 and 12.2 can correspond to observed multiplets. The correspondence with the usual $SO(3)$ spin in the $SO(4, 1)$ representations can be established by reducing the $SO(4, 1)$ representations according to their $SO(4)$ and $SO(3)$ subgroups. This leads to Table 12.4, which shows that Table 12.1 contains only "odd" $SO(4, 1)$ and $SO(3)$ spins.

We may now tentatively classify the particles as follows:

1. **Leptons** (and antileptons) can be classified in the lowest $SO(6, 1)$ representation $\{8\}$ of Table 1 with $B = 0$. On application to our case of Lovelace's proposal [11] (with $\langle Y, Q, L|$ and $|Y, Q, L\rangle$ denoting the usual brackets), let us take $\langle\frac{1}{2}, 1, \frac{3}{2}| = \gamma_5\mu^c$ and $\langle\frac{1}{2}, 1, \frac{1}{2}| = \gamma_5 e^c$, where $e^c = -\gamma_4 Ce^*$, etc. In this choice, μ is a $SU(3)$ singlet; e, $-\nu'$, $\gamma_5\nu^c$ form a unitary triplet, the e, ν, ν' mass differences resulting from the electromagnetic mass breaking. As $8 \times 8 = 35 + 21 + 7 + 1$, we note that only the representation $\{21\}$ corresponds to conserved currents. Under $SU(2) \times SU(2)$ one obtains the familiar vector parts of the conserved weak currents,

Table 12.4. $SO(4, 1)$ representations and their subgroups

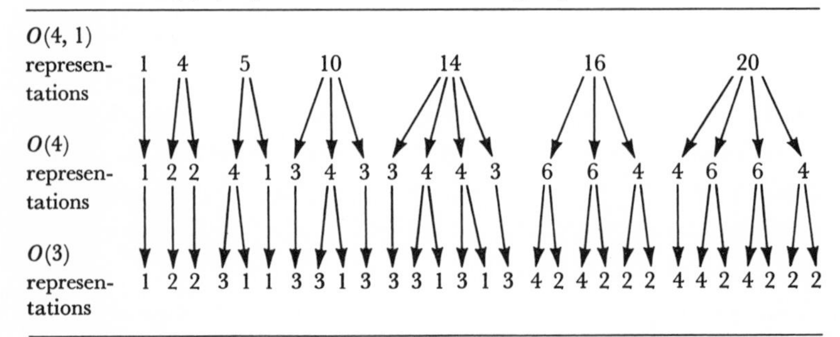

namely $2^{1/2}(\bar{e}\gamma_\mu\nu + \bar{\mu}\gamma_\mu\nu')$ and $2^{1/2}(\bar{\nu}\gamma_\mu e + \bar{\nu}'\gamma_\mu\mu)$, corresponding to $E^{\beta+\gamma}$ and $E^{-\beta-\gamma}$.

2. **Baryons** will be represented in the $SO(6, 1)$ fermion family with $B = \pm 1$. We represent "quarks" by (0, 0, 1)—which can be eliminated physically by an argument such as Komar's[12] or by considering the $SO(6, 1)/Z_2$ group. We thus get:

a. An "octet" family ($L = \pm\frac{3}{2}, \pm\frac{9}{2}, \ldots$) with representations $\{1\}, \{8\}, \{10\}, \ldots$, in which one can introduce, for example,

i. the octets $(\frac{1}{2})^+\{N(939), \Lambda(1115), \Sigma(1190), \Xi(1317)\}$,
$(\frac{5}{2})^+\{N(1688), \Lambda(1815), \Sigma(1880)$ instead of $(1910), \Xi(1930)\}$,
$(\frac{9}{2})^+\{N(2190), \Lambda(2350), \Sigma(2455), \Xi(?)\}$;

ii. the decuplets $(\frac{3}{2})^+\{\Delta(1236), \Sigma(1385), \Xi(1530), \Omega(1672)\}$,
$(\frac{7}{2})^+\{\Delta(1950), \Sigma(2030), \Xi(?), \Omega(?)\}$, $(\frac{11}{2})^+\{\Delta(2420), \Sigma(2650)\}$;

iii. the singlets $\{\Lambda^-(1405), (\frac{1}{2})^-, \Lambda^-(2000?), (\frac{5}{2})^-\}$ and
$\{\Lambda^-(1520), (\frac{3}{2})^-, \Lambda^-(2100), (\frac{7}{2})^-\}$.

b. A "triplet family" ($L = \pm\frac{1}{2}, \pm\frac{5}{2}, \ldots$) with representations $\{3\}, \{6\}, \{15\}$, etc., in which one can introduce, for example,

i. triplets $(\frac{1}{2})^-\{N(1550), \Lambda(1610)\}$, $(\frac{5}{2})^-\{N(2060), \Lambda(2150?)\}$;

ii. triplets $(\frac{3}{2})^-\{N(1518), \Lambda(1650)\}$, $(\frac{7}{2})^-\{N(2190), \Lambda(2350)\}$;

iii. sextets $(\frac{3}{2})^-\{\Sigma(1660), \Xi(1815), \Omega(?)\}$,
$(\frac{7}{2})^-\{\Sigma(2455), \Xi(?), \Omega(?)\}$.

To obtain given $SU(3)$ multiplets with the same parity, within different $SO(6, 1)$ representations, a $J = 2$ jump is necessary. The preceding classification clearly leaves aside the known resonances [such as $\Lambda^-(1640), (\frac{1}{2})^-$, etc.], which can all be regarded as compound states in Maglic's sense,[13] as we shall show later.

3. **Bosons** will be represented in the $SO(6, 1)$ Bose family with $B = 0$. Here too we have:

a. An "octet family" ($L = 0, 3, 6, \ldots$) with $SU(3)$ representations $\{1\}, \{8\}, \{10\}, \{27\}, \ldots$, in which one can put

i. octets $0^-\{M(140), K(495), \eta(545)\}$,
$2^-\{A_2(1270), K(1320), A_2^0(1330?)\}$;

ii. octets $1^-\{\rho(765), K^*(890), \eta'(1019)\}$,
$3^-\{K_sK_s(1410), K(1420), \eta(1600)\}$;

iii. singlets $0^-\{V^o(958), Z^-(1740)\}$, $0^+\{\sigma(410), Z^+(1320)\}$, $1^-\{\omega(787), Z^-(1420)\}$.

b. A triplet family ($L = 1, 3, \ldots$), in which can be fitted the triplets $0^+\{K(720), \varepsilon(730)\}$, $2^+\{K(1420), \eta(1440)\}$, $1^+\{K(1230), D(1285)\}$, $3^+\{K(1780), \eta(1830)\}$.

As in the case of baryons, this does not cover the whole experimental field, and compound states will have to be added. To satisfy experimental data and to give a Yukawa D coupling, the 0^- octet goes over into $\{27\}$ and the 1^- octet into $\{21\}$, along with a M^{++} which decays into $\mu^+ + \mu^+$ and contributes to the $e - \mu$ mass breaking.

We now calculate the mass formula. In our model, $\mu^2 = C_1 = M_{\alpha\mu}M^{\alpha\beta}$ ($\alpha, \beta = 0, 1, 2, 3, 4$) is the first Casimer operator of g and g'. The group $SO(6, 1)$ (rank 3, with 21 generators) permits twelve commuting simultaneously diagonalizable operators. $G = U(1) \times SO(6, 1)$ thus permits thirteen diagonalizable operators, $U(1)$ corresponding to the unit matrix in any finite representation. In the basic representation one can define seven matrices satisfying the Clifford algebra $(\overline{\Gamma}_\alpha, \overline{\Gamma}_\beta)_+ = 2g_{\alpha\beta}(\alpha, \beta = 0, 1, 2, 3, 4, 5, 6)$. The dimensions must be doubled twice (to get 32) if one wants to introduce right and left translations of $SO(6, 1)$ and the discrete automorphisms such as PCT. Now, in any finite representation we can diagonalize: (1) the $U(1)$ B unit matrix; (2) the three Casimers of $SO(6, 1)$; (3) the three H_α generators Y, T_3, L (or Y, Q, L); (4) three supplementary generators of $SU(3)$, viz. $T(T + 1)$ and the two $SU(3)$ Casimer operators; (5) three supplementary operators of $SO(4, 1)$, namely two Casimers (which include μ^2) and two more, since we have six diagonalizable operators in all for $SO(4, 1)$. The total number of operators being fourteen, one must impose a polynomial constraint among them, as $U(1) \times SO(6, 1)$ is semisimple. If we write these operators as A_σ ($\sigma = 0, 1, \ldots, 12$), the constraint becomes

$$\sum_\sigma a_{1\sigma}A_\sigma + \sum_\sigma a_{2\sigma}A_\sigma^2 + \cdots = \text{const.},$$

which can be transformed into the mass formula

$$\mu^2 = a_0' + \sum_{\sigma'} a_{1\sigma'} A_{\sigma'} + \sum_{\sigma'} a_{2\sigma'} A_{\sigma'}^2 + \cdots,$$

where $\sigma' = 1, 2, \ldots, 12$. The $a_{i\sigma}$ and $a_{i\sigma'}$ are constant scalars or pseudoscalars under the E discrete automorphisms; these coefficients clearly depend on the chosen representation and its normalization. Indeed, denoting by E any E^α generator of $SO(6, 1)$ which induces transitions within $SU(3)$ multiplets, and by $|a\rangle$ and $|b\rangle$ two $SU(3)$ states, we get $E|a\rangle = e_{ab}|b\rangle$ (where e_{ab} is a matrix element), $\mu^2|a\rangle = m_a^2|a\rangle$, and $\mu^2|b\rangle = m_b^2|b\rangle$, so that

$$\langle b|(\mu^2, E)|a\rangle = (m_b^2 - m_a^2)\langle b|b\rangle e_{ba}.$$

The above mass relation can thus be made more explicit. Its first-order terms, which contribute to the mass difference within any given $SU(3)$ multiplet, take the form $aY + bT(T + 1)$ Similar multiplets within different $SO(6, 1)$ representations are separated by a term cn (where n denotes the number of quarks) which results from the $U(1)$ contribution, since the $U(1)$ contribution in $\{8\}$ assumes the form $c(PC)B$, where $c(PC)$ is a pseudoscalar under PC. In this case, since the basic quarks have $J = \frac{1}{2}$, the $U(1)$ contribution to the mass formula is cJ, as the total J is the sum of the quark J's. For a given $SU(3)$ multiplet we get, neglecting higher-order terms (which can be eliminated, as we will show elsewhere, by a closer examination of interactions),

$$\mu^2 \approx m_0^2 + aY + bT(T + 1) + cJ;$$

here we have absorbed into m_0^2 any contribution from first-order terms that do not vary for a given $SU(3)$ multiplet. This formula, earlier discovered for other groups within unification theory by Flato et al.,[14] fits astonishingly well all known baryon and boson resonances, with the exception of resonances which cannot be regarded as baryon-boson or boson-boron compound states, and it recovers all the results of Regge-pole theory. Moreover, it is seen that the coefficients m_0, a, b, c in our formula are not independent for baryons and bosons. Indeed, since "true" resonances appear in this picture as internal discrete quantized states, we must adopt Maglic's phenomenological point of view and consider: (a) that quantum jumps between baryons produce heavy boson quanta, i.e. $\Delta m(\text{baryons}) \cong \Sigma m(\text{bosons})$, and (b) that quantum jumps between bosons produce bosons, i.e. $\Delta m(\text{bosons}) \cong \Sigma m(\text{bosons})$; so that the coefficients for baryons determine those for bosons.

Two remarks are appropriate in conclusion. First, one must use in external motions the infinite unitary representations of $SO(4, 1)$ induced in Wigner's manner from the corresponding finite ones employed in the internal $SO(6, 1)$ representation to classify the particles' internal states. One could also introduce infinite unitary representations both of $SO(6, 1)$ and $SO(4, 1)$. This is mathematically possible, because, as shown by Ottoson,[15] all $SO(p, 1)$ unitary representations are explicitly calculable as infinite direct sums of finite $SO(p)$ representations. Second, we note that $SO(6, 1)$ has an internal total phase $T_3 + Y/2 + L/3$ which should be equated to the external phase of $U(1) \times SO(4, 1)$ on the particle's boundary to ensure a suitable fitting.[16] This equalization yields a simple physico-geometrical interpretation of the Nishijima–Gell-Mann formula.

The author would like to thank Professors L. de Broglie, N. N. Bogolubov, C. Lovelace, and M. Flato for helpful discussions. He is also grateful to M. Zulauf, P. Gueret, and S. Depoquit for assistance in examining group-theoretical and physical aspects of the aforementioned model.

References

1.
M. Flato, D. Sternheimer, and J.-P. Vigier, Compt. Rend. **260**, 3869 (1965).
2.
M. Flato, *Thesis* (Gauthier Villars, 1965).
3.
L. O'Raifeartaigh, Phys. Rev. **139**, 1052 (1965); Phys. Rev. Letters **14**, 332 (1965).
4.
M. Flato and D. Sternheimer, Phys. Rev. Letters **15**, 934 (1965).

5.
Refs. 1 and 2.
6.
L. de Broglie, D. Bohm, P. Hillion, F. Halbwachs, T. Takabayasi, and J.-P. Vigier, *Proc. Aix Conf.* **1**, 503 (1961); Phys. Rev. **129**, 438 (1963); L. de Broglie, F. Halbwachs, P. Hillion, T. Takabayasi, J.-P. Vigier, Phys. Rev. **129**, 451 (1963).
7.
M. Galto, CERN Preprint (1968).
8.
J.-M. Souriau, Nuovo Cimento **30**, 565 (1963).
9.
C. Lovelace, Nuovo Cimento **37**, 225 (1965).
10.
E. B. Dynkin, Am. Math. Soc. Translations **6**, 111, 145 (1957).
11.
Ref. 9.
12.
M. Flato and J. Sternheimer, Compt. Rend. **259**, 3455 (1964).
13.
B. C. Maglic, Nuovo Cimento **45**, 949 (1966).
14.
D. Bohm, M. Flato, F. Halbwachs, P. Hillion, and J.-P. Vigier, Nuovo Cimento **36**, 672 (1965).
15.
U. Ottoson, Preprint, Institute of Theoretical Physics, Göteberg, Sweden.
16.
J.-P. Vigier, Phys. Rev. Letters **17**, 39 (1966).

Thirteen **Karl R. Popper**

Particle Annihilation and the Argument of Einstein, Podolsky, and Rosen

It may appear as if all such considerations were just superfluous learned hairsplitting. . . . However, it depends precisely upon such considerations in which direction one . . . must look for the future . . . basis of physics.

Albert Einstein[1]

1. Introduction

Alfred Landé's fight for clear thinking in quantum mechanics consists in the main of two great efforts: in a criticism of the orthodox interpretation (Copenhagen interpretation) of the formalism and in a reconstruction of the theory as a straightforward statistical theory of particles. I do not doubt, however, that Landé never suggested that even a rational interpretation of the formalism of quantum mechanics would be the last word in the theory of matter. In spite of its tremendous success, quantum mechanics suffers from a number of ills. Some of these Landé could remove, but some important ones (such as the divergences) remain. And even if we forget about some inconsistencies, the theory is, to use Einstein's famous indictment, "incomplete." In fact, it seems "incomplete" in more than one sense of this term, as will emerge later.

Thus a fight for clear thinking in quantum mechanics, like Landé's, becomes also a fight for the removal of obstacles to new creative thinking, to the invention of new theories. I regard this as very important. Einstein was convinced of the need for a new basis of physics. His opponents, the upholders of the Copenhagen interpretation, sometimes said that they too were looking for a new basis, although they were looking in a different direction. Nobody, least of all Einstein, would have denied their right to do so.

I am greatly indebted to Professor Abner Shimony for his comments on an earlier version of this paper.

But they did complain that Einstein was not convinced by the (extremely vague) argument by which they tried to show that the quantum-mechanical revolution was final. For example, Bohr quotes in 1949 the following remark of Einstein's on the view (Bohr's view) that quantum mechanics offers an exhaustive description of individual phenomena: "To believe this is logically possible without contradiction; but it is so very contrary to my scientific instinct that I cannot forgo the search for a more complete conception."[2]

But instead of appreciating Einstein's extremely conciliatory attitude, and letting him go his own way, Bohr comments on Einstein's remark: "Even if such an attitude might seem well-balanced in itself, it nevertheless implies a rejection of the whole argumentation"—that is, Bohr's argumentation that aimed to show "that, in quantum mechanics, we are not [confronted] with an arbitrary renunciation of a more detailed analysis of atomic phenomena, but with a recognition that such an analysis is *in principle* excluded."

Bohr, it seems, thought that his argumentation amounted to a conclusive proof that every reasonable person had to accept and only an unbalanced person could possibly reject. But quite apart from the vagueness of Bohr's argumentation, *such proofs do not exist*: in science we are always guessing. This does not mean that the point of view for which Bohr tried to argue could not turn out to be the right one: Einstein admitted this possibility. But the fact that Bohr's argumentation does not amount to a proof means that Einstein had as much right to guess as had Bohr, and without having it questioned whether his "attitude" was "well-balanced" or not.

Quantum mechanics is guesswork, like every other physical theory. Its success is as impressive as that of Newton's theory of gravitation in, say, 1915. But while Newton's theory never got into very serious trouble, quantum mechanics has done so. It is worth remembering that a theory of the constitution of matter that has nothing to say, by way of explanation, about the electronic charge (though it looks very much like a quantum effect), or about the ratio between the masses of electrons and protons, cannot be regarded as "complete"; and the tremendous success of the theory in explaining the periodic table of the elements, and more recently in developing a theory of the nucleus, is somewhat balanced by its inability to explain the superabundant elementary particles. Also, there is so far no satisfactory link between quantum mechanics and gravitation. All this points to the need for a more general theory; and what Einstein asked for was the freedom to link his guesses to a criticism of quantum mechanics.

Landé's exposures of the at times really atrocious utterances of the upholders of the Copenhagen interpretation is an implicit criticism of their claim to understand physics so much better than everyone else that they can pronounce on and choose the way into its future.

Fortunately there have always been critics. Dirac spoke very clearly at the end of his *Principles* about the unforeseeable character of a new revolution.[3] Schrödinger was never satisfied and always hoped for a new development. De Broglie has outlined a theory that is most promising. There was Temple's paradox[4] (based on von Neumann's theory[5]), which has recently been resuscitated by Park and Margenau[6] (see also my note in *Nature* of 1968[7]). There were Margenau,[8,9] Bohm,[10] Vigier,[11,12] and Bunge.[13] And recently there have been Bell[14–16] and Nelson.[17,18]

It so happened that I too was engaged in this field at an early date, although I was, and remained, an outsider: I published in 1934 a mistaken thought experiment, first in a short note,[19] and then in a section of a book.[20] The fact that this thought experiment contained a mistake discouraged me for years from any further publication in this field, and it was largely Alfred Landé's encouragement that broke this spell.

Yet my mistaken thought experiment was only part of a larger critical argument against the Copenhagen interpretation and against its attempts to issue "prohibitions that draw limits to the possibilities of [future] researches."[21,22] Also, my mistaken thought experiment was, in a way, a (very clumsy) precursor of the thought experiment of Einstein, Podolsky, and Rosen.[23,24] Like the latter experiment, mine also operated with *two particles (or systems) that had interacted*, and also with the idea that in measuring one of them (S_1, say) *we can obtain information concerning the other* (S_2).

My short note in *Die Naturwissenschaften* of 1934 was successfully criticized in the same number by von Weizsäcker,[25] and there was an exchange of private letters over it, with him and also with Heisenberg. Yet I was doubtful about these refutations until I received in 1935 a letter from Einstein that also contained a refutation, and later another one[26] that, besides a more detailed refutation, contained an outline of the now famous argument of Einstein, Podolsky, and Rosen (E-P-R).[27] These convinced me of my mistake. Later in the same year (1935) I had an opportunity to discuss the matter with Schrödinger, who was deeply interested in E-P-R,[28] and who did not believe in the adequacy of Bohr's reply[29] any more than Einstein did. In the summer of 1936 I met Bohr (it was Victor Weisskopf who introduced me to him). In several prolonged discussions with Bohr these matters were taken up, and I was completely overwhelmed by his irresistible charm and vigor. For a time I gave up the attempt to understand; but I was never quite convinced by Bohr's reply to E-P-R, and for years I was in despair about my failure to understand Bohr's complementarity and his interpretation of quantum theory.

The struggle between Einstein and Bohr was taken up again after the war, in 1948, in an issue of *Dialectica* edited by Wolfgang Pauli.[30] The contributions of Einstein[31] and of Bohr[32] to this number were, I still think, more pointed and explicit than their earlier contributions.[33] Bohr stressed very clearly the need of taking the *total experimental situation* into account. Einstein, on the other hand, stressed the old principle of *local action* ("no action at a distance") or, as it is now called by Bell,[34] the principle of locality. In 1950 I met Einstein and Bohr in Princeton. At the time my main concern was the thesis that not only quantum mechanics but also classical physics was indeterministic, in spite of the "*prima facie* deterministic character" of classical mechanics (see a paper of mine of 1950[35]). This is a thesis which Alfred Landé put in 1953 in a very much clearer and better way.[36] Later I found that C. S. Peirce had held similar views, and also Franz Exner in Vienna, as reported by Schrödinger.[37,38] More recently, the thesis has become fashionable. But I think that the strongest argument in its favor is still the one due to Landé.

At about this time I read David Bohm's beautiful *Quantum Theory.*[39] There E-P-R was taken very seriously and discussed on what I believe are radically new lines, on which I will now comment.

2. Bohm's Variant: E-P-R and E-P-R-B Compared

In his *Quantum Theory*[40] David Bohm suggested a variant of E-P-R that has proved immensely important. He suggested that the measurement of the position and momentum coordinates of the two particles or subsystems, S_1 and S_2, described by E-P-R, should be replaced by a measurement of freely selected components of the spins of S_1 and S_2 after their separation. Alternatively, we could measure the states of polarization of S_1 and S_2, provided S_1 and S_2 are photons. In what follows I will, for the sake of simplicity, assume that S_1 and S_2 are photons and that we are "measuring" with the help of a polarizer the state of polarization of S_1. As pointed out by Bohm, we can then calculate, by the E-P-R argument, the state of polarization of S_2. I will call this the E-P-R-B experiment.

I will first explain the reason why the E-P-R-B experiment makes so important a difference to the E-P-R argument. (I call it an "argument" rather than a "paradox"; compare, for example, Yourgrau[41] and my comments.[42])

Einstein, especially in his later discussions[43,44] of the E-P-R argument, always stressed that two interpretations of the situation are possible:

(1) The systems (particles) S_1 and S_2 possess, *before* they are measured, definite values of the variable q (or alternatively p),

and the measurement of (say) q_1 merely determines a pre-existing value; that is to say, the measurement of q_1 yields the value of q_1 which was a characteristic property of S_1 even before it was measured.
(2) S_1 has no definite value of q_1 before the measurement. The value of q_1 which results from the measurement comes into being when S_1 is being measured.

Einstein opted for (1), not because this was the "classical" view of measurement, but for the following decisive reason.

If (2) holds, and if a measurement of S_1 is undertaken at a time when S_1 and S_2 are in widely separated regions of space, then any (maximal) measurement of S_1 would mean, from the point of view of quantum mechanics, (a) the separation of the system $(S_1 + S_2)$ into two systems, S_1 and S_2, which from that moment on are no longer dependent on each other, and (b) if the variable q_1 of S_1 is measured, then the variable q_2 of S_2 becomes definite, and the variable p_2 becomes (intrinsically) indefinite (or "smeared"); and if the variable p_1 of S_1 is measured, the variable p_2 of S_2 becomes definite, and q_2 becomes intrinsically indefinite (or "smeared"). But if S_2 is far away from S_1 at the time S_1 is being measured, then there will be an action at a distance exerted from the measurement of S_1 upon the system (or the particle) S_2.

Thus if we adopt the principle of local action, we have, it appears, to reject (2) and to accept (1).

As Einstein made clear[45] Bohr accepted (2), even though he did not explicitly mention action at a distance. (See in this connection expecially Bohr's paper in *Dialectica*,[46] which, as it emerges from Pauli,[47] was written after Einstein's paper in the same number.[48] Bohr's famous contribution to the volume *Albert Einstein, Philosopher-Scientist*[49] was written later still, as emerges from the paper itself.[50])

Now Bohm's new argument, E-P-R-B, amounts to offering a case in which Einstein's interpretation (1) is ruled out from the very beginning. For nobody assumes that if we "measure" a state of polarization of some system S (it does not matter whether we think in terms of waves or photons) with the help of a polarizer, we always measure the state of S as it was immediately before entering the polarizer; on the contrary, we may actually know that the state of polarization of S before entering the polarizer was different from the state "measured" by this polarizer. (In this case we are faced not with a mere "measurement" of the state of S but with an interaction that changes the known state of polarization of S into another known state.)

Accordingly, Einstein's formulation of interpretation (1) has to be replaced for the E-P-R-B case by something like the following (1′):

(1′) The systems (particles) S_1 and S_2 possess before the measurement definite values of the variable to be measured; but the result of a "measurement" may not coincide with these pre-existing values.

Now this change from (1) to (1′) is of the greatest significance in connection with Einstein's argument from the principle of local action. For if (as in the E-P-R-B case) the "measurement" of S_1 consists in S_1 passing through a polarizer, and if *this* "measurement" of S_1 informs us according to quantum mechanics about the state of S_2, then the kind of action at a distance described by Einstein is not merely part of *interpretation* (2)—that is, of Bohr's interpretation of quantum mechanics—but part of quantum mechanics itself.

This, I think, is the great difference which Bohm's suggestion makes to the E-P-R argument. It means, in a sense, the refutation of Einstein's argument, since he did not use it to argue against the truth of a quantum-mechanical assertion but merely against the completeness of quantum mechanics. But I do not doubt that Einstein would have upheld his argument against action at a distance and would have asserted that since quantum mechanics asserts action at a distance in the sense described, it is likely to be false. (I may perhaps say here that I would find it unprofitable to discuss the question whether the situation described is properly to be denoted by the term "action at a distance": it is admitted by all sides—certainly by Einstein—that this kind of action at a distance cannot be used to transmit signals with a velocity faster than light, but the same holds for Newton's action at a distance; and nobody asserts that such action at a distance is inconceivable. Einstein's argument is merely that he does not believe that the physical world has this kind of character, call it by any name you like.)

This would lead to the view that if we can design an actual E-P-R-B experiment, it might be a crucial experiment to decide between a quantum-mechanical "action at a distance" and Einstein's principle of local action.

3. Actual Experiments: Wu and Shaknov; Kocher and Commins; Clauser and His Collaborators

The original E-P-R thought experiment has long ceased to be a mere thought experiment. With particle annihilation that leads to the *creation of twin photons*, it has become a real experiment.

We assume an arrangement in which a (spatially very small) source of twin photons (positronium surrounded by some material arresting the positrons) is placed at the origin of our coordinate system. We place a (small) gamma ray detector on, say, the negative x-axis; the latter may be taken to be defined by the source together with the detector.

Now if a gamma ray S_1 registers in our detector, then we know that another gamma ray S_2 is on its way along the positive x-axis and that S_2 has reached the same distance from the origin as S_1. On the other hand, we can measure the hardness of S_1 and thereby its momentum, and thus obtain that of S_2. These predictions for S_2 can be tested in their turn by appropriate measurement. (This is why I wrote: "The Einstein-Podolsky-Rosen experiment has since become a real experiment, in connection with pair creation, and [with] pair destruction with photon-pair creation."[51])

I think that Einstein's original argument fully applies to this case: the fact that there also exist "measurements" of a completely different type (such as polarizers) need not deter us from applying his original argument to this case. In fact, I possess somewhere a letter of Schrödinger's in which he says, of a similar case of position measurement, that "only a congenital idiot" would think that in such a case the position was not there before but is the result of the process of measurement. This is perhaps a bit strong, but it reminds us that we should not surrender common sense too easily.

In principle, twin photon creation may be used to turn even E-P-R-B into a real experiment. This has been done by Wu and Shaknov.[52]

In Wu and Shaknov's experiment, an attempt is made to measure the states of polarization of the twin photons, S_1 and S_2, that are created by particle annihilation. The trouble with this experiment is that there are no polarizers for γ-rays. Otherwise the experiment would be a crucial experiment between quantum-mechanical action at a distance and Einstein's insistence on local action. For the statistical predictions between these two theories differ: quantum mechanics predicts that the coincidence rate for two detectors (detecting S_1 and the coinciding S_2) will be cut to $\frac{1}{2}$ by placing polarizers in crossed positions before the detectors, and to 0 if the polarizers are in parallel positions. Einstein's view (sometimes characterized by the term "hidden variable theory") would lead, according to Frisch,[53] to the ratios $\frac{3}{8}$ and $\frac{1}{8}$ respectively.

According to Frisch[54] (see also Dicke and Wittke[55]), Wu and Shaknov's result can be interpreted as confirming quantum mechanics. However, in a recent paper by Clauser, Horne, Shimony, and Holt[56] it is pointed out that the methods used by Wu and Shaknov cannot lead to a decisive result. (See also Bohm and Aharonov.[57]) The reason is that in the absence of polarizers, Wu and Shaknov had to use indirect statistical methods (Compton scattering) for determining states of polarization. But, as Clauser and his co-workers point out, "The direction of Compton scattering of a photon is a statistically weak index of its linear polarization."

For this reason* the authors do not regard the results of Wu and Shaknov as sufficient to inform us about the state of polarization as predicted by quantum mechanics on the one side and by a "hidden variable theory" on the other. Looking out for an arrangement permitting the use of polarizers, they found that the experiment of Kocher and Commins[58] could be adapted so as to become crucially decisive between standard quantum theory and all "local hidden variable" theories, that is to say, all hidden variable theories satisfying the principle of local action (no action at a distance). The experiment of Kocher and Commins permits the use of polarizers (because here the twin photons are not produced by particle annihilation and so are sufficiently soft), and the statistics could therefore be relied on to determine simply the probabilistic effect upon each pair of twin photons of the (crossed, parallel, or intermediary) positions of the two polarizers.

4. Comments on the View of Bell

Bohr's original reply to E-P-R was very difficult to understand, but it became clearer in the form given to it by Furry.[59] Its main point, it seems, was that the total experimental arrangement had to be assumed as given in advance. Consequently results in some part of the experimental setup may then appear to be dependent upon parts of it which are locally remote. (In this view E-P-R would become something like an extreme case of the two-slit experiment —that is, with the slits at a great distance.) Now this would mean,

* From the point of view of Einstein's distinction between (1) and (2) (as discussed above), Compton scattering (assuming it could furnish a sufficiently strong index of polarization) might be interpreted in the sense of (1) rather than (1′): as opposed to "measurement" by a polarizer, Compton scattering may be interpreted as measuring a pre-existing state. If this point of view were adopted, then Compton scattering could not be used for a crucial test: even a perfect agreement between results of the Wu and Shaknov type and the predictions of quantum mechanics could not refute a "local hidden variable" interpretation in the sense of Bell. (Einstein's distinction between (1) and (2), and our distinction between (1) and (1′) are of course unacceptable for the quantum orthodoxy although the thought of "correspondence" arguments might give them pause.)

in the E-P-R case, that we *either* have to assume, as pointed out by Einstein, some instantaneous action at a distance, from one part of the experimental setup on another part, or else we assume that the experimental setup would have to be fixed sufficiently in advance of the experiment, so as to allow its parts to establish (by some local action) a kind of "mutual rapport," to use the terminology of J. Bell.[60] Bell distinguishes these two possibilities and very clearly states the second possibility as follows: "quantum mechanical predictions . . . might apply only to experiments in which the setting of the instruments are made sufficiently in advance to allow [the instruments] to reach some mutual rapport by exchange of signals with velocity less than or equal to that of light."[61] I propose to call this very interesting second possibility "Bell's conjecture of advance rapport" or more briefly "Bell's rapport conjecture."

Bell's rapport conjecture seems at least partially to agree with some of Bohr's ideas, especially with Bohr's stress upon the significance of the total experimental setup. Yet it would mean that quantum mechanics would not fully apply to cases in which the experimental arrangements are not prearranged or completed "sufficiently in advance." Thus quantum mechanics would turn out to be a theory "of limited validity," as Bell puts it; indeed, as a theory that claims universal validity, it would simply be false. In Einstein's terminology, it also could not be "considered complete," because it would not explain how the "rapport" is established. (There would be "hidden variables" by which to establish rapport.)

Thus we have to distinguish between *three* theories:

(1) Standard quantum mechanics that, it turns out, implies action at a distance (in Einstein's sense).

(2) Quantum mechanics with "limited validity" in the sense of Bell's rapport conjecture. This would mean that quantum mechanics is only approximately true (i.e. that it is false) and also that it is incomplete.

(3) A purely statistical interpretation of quantum mechanics on the lines sometimes envisaged by Einstein or Landé or de Broglie

(and also by myself). In the terminology of Bell, this would involve the thesis of the existence of "local hidden variables"; and it would not (in view of E-P-R-B) be fully consistent with quantum mechanics: there would be crucial experiments.

It should be noted—and this seems to me important—that the experiment recently proposed by Clauser, Horn, Shimony, and Holt,[62] would be crucial only between (1) and (2), on the one hand, and (3), on the other: it cannot decide between (1) and (2). In order to decide between (1) and (2), we should need "experiments of the type proposed by Bohm and Aharonov, in which the settings of the polarizers are changed during the flight of the particles." (Bell.[63])

It is not the purpose of this paper to discuss the problem of "hidden variables," and even less to find flaws in Bell's most impressive work.[64] Yet the following points might be noted in passing. If correct, they may lead to a reinterpretation of Bell's proof of the incompatibility of quantum mechanics with the assumption of local "hidden variables"—a reinterpretation that seems to differ both from Bell's own interpretation and from that adopted in the paper by Clauser et al.[65] It would also lead to a simplification of both Bell's proof and the logic of the situation.

When first reading Bell,[66] I thought that his very general proof—that the assumption of "local hidden variables" of any kind is incompatible with quantum mechanics—did not make allowance for (local) stochastic hidden variables but only for deterministic ones. Although this appears to be so for his proof *as it stands*, it nevertheless seems to me that the proof can be, trivially, extended to the stochastic case for the following reason. Standard quantum mechanics, i.e. the theory (1), implies action at a distance in Einstein's sense. But this obviously clashes with "locality" in Bell's sense. Thus we do not need to revise Bell's ingenious and detailed proof.

Further, it should be noted (and this seems to me of considerable importance, even though it is quite obvious) that if quantum mechanics is modified so as to make allowance for Bell's rapport conjecture, then for the so modified theory Bell's proof of the incompatibility between quantum mechanics and local "hidden variable theories" no longer holds. For this proof makes essential use of the fact that quantum mechanics implies some kind of action at a distance.

5. An Afterthought

We may speculate about what would have happened in 1927 if a straightforward and satisfactory classical statistical derivation of the Schrödinger equation (such as Edward Nelson's[67]) from a classical statistical model had been available. My guess is that complementarity and the Copenhagen interpretation would never have arisen: there would have been no need for anything of the kind.

I am aware of the fact that the Schrödinger equation is not everything. But it was almost everything in 1927. What goes beyond it did come later; and it all came as a surprise.

References

1.
P. A. Schilpp, ed., *Albert Einstein, Philosopher-Scientist* (Library of Living Philosophers, 1949), p. 683.
2.
Ibid., p. 235.
3.
P. A. M. Dirac, *The Principles of Quantum Mechanics*, 4th ed. (Clarendon Press, 1938).
4.
G. Temple, Nature **135**, 957 (1935); and **136**, 179 (1935).
5.
J. von Neumann, *Mathematical Foundations of Quantum Mechanics* (Princeton University Press; 1949, 1955; German ed. 1932).
6.
J. Park and H. Margenau, Intern. J. Theoret. Phys. **1**, 211 (1938).
7.
K. R. Popper, Nature **219**, 682 (1968).
8.
H. Margenau, *The Nature of Physical Reality* (McGraw-Hill, 1950).

9.
H. Margenau and L. Cohen, in *Quantum Theory and Reality*, ed. by M. Bunge (Springer-Verlag, 1967).
10.
D. Bohm, Phys. Rev. **85**, 166 and 180 (1952).
11.
J.-P. Vigier, in *Observation and Interpretation*, ed. by S. Körner (Dover, 1957).
12.
J.-P. Vigier, in *Physics, Logic, and History*, ed. by W. Yourgrau and A. D. Breck (Plenum Press, 1970), pp. 191–202.
13.
M. Bunge, Brit. J. Phil. Sci. **6**, 1 and 141 (1955).
14.
J. S. Bell, Physics **1**, 195 (1964).
15.
J. S. Bell, Rev. Mod. Phys. **38**, 447 (1966).
16.
J. S. Bell and M. Nauenberg, in *Preludes in Theoretical Physics*. In Honor of Victor Weisskopf, ed. by A. De-Shalit ,H. Feschbach, and L. Van Hove (North-Holland, 1966).
17.
E. Nelson, *Phys. Rev.* **150**, 1079 (1966).
18.
E. Nelson, *Dynamic Theories of Brownian Motion* (Princeton University Press, 1967).
19.
K. R. Popper, Naturwissenschaften **22**, 807 (1934).
20.
K. R. Popper, *Logik der Forschung* (J. Springer-Verlag, 1934), Sect. 77 (3d ed., 1969).
21.
Ibid.
22.
K. R. Popper, *The Logic of Scientific Discovery* (Basic Books 1959, 1968), end of sec. 78.
23.
A. Einstein, B. Podolsky, and N. Rosen, Phys. Rev. **47**, 777 (1935).
24.
A. Einstein, A letter dated 1935, in K. Popper, *Logic of Scientific Discovery*.
25.
C. F. von Weizsäcker, Naturwissenschaften **22**, 808 (1934).
26.
Ref. 24.
27.
Ref. 23.
28.
E. Schrödinger, Proc. Cambridge Phil. Soc. **31**, 555 (1935).

29.
N. Bohr, Phys. Rev. **48**, 696 (1935).
30.
W. Pauli, ed., Special issue (No. 7/8) of Dialectica **2**, No. 3/4, 307–422 (1948).
31.
A. Einstein, in Ref. 30.
32.
N. Bohr, in Ref. 30.
33.
Refs. 23 and 29.
34.
Refs. 14 and 15.
35.
K. R. Popper, Brit. J. Phil. Sci. **1**, 117 and 173 (1950).
36.
A. Landé, *Foundations of Quantum Theory* (Yale University Press, 1955), pp. 3ff.
37.
E. Schrödinger, Naturwissenschaften, Sept. 17, 1929. (English translation, Chap. 6 of Ref. 38.)
38.
E. Schrödinger, *Science, Theory and Man* (Norton, 1935; McLelland, 1957), pp. 71 and 142.
39.
D. Bohm, *Quantum Theory* (Prentice-Hall, 1951).
40.
Ibid., p. 614.
41.
W. Yourgrau, in *Problems in the Philosophy of Science*, ed. by I. Lakatos and A. Musgrave (North-Holland, 1968), pp. 182ff.
42.
K. R. Popper, in *Problems in the Philosophy of Science*, ed. by I. Lakatos and A. Musgrave (North-Holland, 1968), pp. 202ff.
43.
Ref. 31.
44.
Ref. 1, pp. 82ff. and 682.
45.
Ibid., pp. 681ff.
46.
Ref. 32.
47.
W. Pauli, Editorial, in Ref. 30, p. 308, last 4 ll.
48.
Ref. 31.
49.
N. Bohr, in Ref. 1.

50.
Ibid., p. 239.
51.
M. Bunge, ed., *Quantum Theory and Reality* (Springer-Verlag, 1967), p. 28.
52.
C. S. Wu and I. Shaknov, Phys. Rev. **77**, 136 (1950).
53.
(Added in proofs.) See Ref. 54. Professor Leslie E. Ballentine, who has read the present paper in proofs, questions this result, referring to Bell's results according to which there *can* be local hidden variable theories that "agree with Q. M. for certain relative angles of the detectors (polarizers), including *parallel* and *perpendicular*."
54.
O. R. Frisch, in *The Critical Approach to Science and Philosophy*. In Honor of Karl Popper, ed. by M. Bunge (Free Press, 1964).
55.
R. H. Dicke and J. P. Wittke, *Introduction to Quantum Mechanics* (Addison-Wesley, 1960).
56.
F. Clauser, M. A. Horne, A. Shimony, and R. A. Holt, Phys. Rev. Letters **23**, 880 (Oct. 13, 1969).
57.
D. Bohm and Y. Aharonov, Phys. Rev. **108**, 1070 (1957).
58.
C. A. Kocher and E. D. Commins, Phys. Rev. Letters **18**, 575 (1967).
59.
W. H. Furry, Phys. Rev. **49**, 393 (1936).
60.
Ref. 14.
61.
Ibid., p. 199.
62.
Ref. 56.
63.
Ref. 14, p. 199.
64.
Refs. 14 and 15.
65.
Ref. 56.
66.
Ref. 15.
67.
Ref. 18.

The following references are relevant although not cited in the article.
68.
N. R. Hanson, *The Concept of the Positron* (Cambridge University Press, 1963).

69.
A. Landé, *Principles of Quantum Mechanics* (Macmillan, 1937).
70.
A. Landé, *Quantum Mechanics* (Pitman Publishing Corp., 1951).
71.
A. Landé, *From Dualism to Unity in Quantum Physics* (Cambridge University Press, 1960).
72.
A. Landé, *New Foundations of Quantum Mechanics* (Cambridge University Press, 1965).
73.
A. Landé, Phys. Today **20**, 55 (1967).
74.
R. Peierls, Nature **136**, 395 (1935).
75.
K. R. Popper, in *Observation and Interpretation*, ed. by S. Körner.
76.
A. Shimony, Phys. Today **19**, 85 (1966).
77.
J. A. Wheeler, Ann. N.Y. Acad. Sci. **48**, 219 (1946).
78.
J. A. Wheeler and R. P. Feynman, Rev. Mod. Phys. **17**, 157 (1945); and **21**, 425 (1949).
(Further references to relevant publications will be found especially in 6, 15, 18, 22 (app. * xi), 52, 56, 58, 72, and 78.)

Fourteen **Walter M. Elsasser**

Philosophical Dissonances in Quantum Mechanics

The endless stream of papers, symposia, and discussions on what is called "the quantum theory of measurement" indicates a deep-seated feeling of discomfort among physicists to the effect that quantum mechanics cannot yet be considered "finished," that even in its more conventional considerations (i.e. outside the more difficult realm of nuclear structure and high energies) there is still some element lacking, that some building stones have been left out. Here we shall agree with those who believe that the monumental success of quantum theory in so many fields, including chemistry, analysis of the solid state, etc., means that the basic *mathematical* apparatus is with us to stay, and so is the statistical interpretation in one form or the other, implying that one cannot return some day to an underlying "deterministic" theory in the sense suggested by Einstein, de Broglie, and so many others of the older generation. But again, we shall have to explain later that the "subjectivist" interpretation of quantum theory was a makeshift which, having done its service, must eventually disappear.

The current state of affairs may be alternately illuminated on recalling the suddenness with which quantum mechanics appeared on the scene and was then developed with a speed seemingly unique in the history of intellectual discoveries. In the year 1900 Planck introduced the quantum of action, and it was in the same year that Ernst Mach, looking through a microscope at the scintillations of α-particles just discovered by Becquerel, withdrew his philosophical objections against speaking of atoms as components of the real world rather than as mere verbal symbols expressing the content of chemical equations. A third of a century later there appeared two pivotal contributions: von Neumann's great work on the mathematical foundations of quantum mechanics and Niels Bohr's basic paper on generalized complementarity. It is fair enough to say that in the period between these signposts practically the whole of quantum theory as we know it today (and so far as it refers to energies below the nuclear level) was conceived and executed, and this to a degree of consistency which was very far from rough.

I was an eyewitness to the last part of this evolution and the following item of personal reminiscence might throw light on the special circumstances which surrounded this sudden growth of quantum theory. The incident happened in 1926 when I was a student of physics at the University of Göttingen. There was a seminar entitled "Structure of Matter" which had been running for years and in which one of the main personages was the celebrated mathematician David Hilbert. Although no longer precisely young, Hilbert attended on a variety of occasions. Once, during a lengthy discussion, he told glowingly about the rapid progress he had made, collaborating with John von Neumann, in applying linear operator calculus to quantum mechanics. He then remarked that this was not the first time he was engaged in such an endeavor, but now it seemed to be definitely successful. Much earlier, around the turn of the century, he had joined together with a physicist (and if my memory does not deceive me at this distance, it was Rydberg, the spectroscopist) when the endeavor had been to find an eigenvalue problem of a linear operator whose characteristic values would be representative of spectral lines. The undertaking failed and had been abandoned for quite specific reasons: Hilbert was able to prove that, given a linear problem with an infinite sequence of discrete characteristic values, say $E_1, E_2, \ldots$, their point of convergence must necessarily be either $E_n \to 0$, or $E_n \to \infty$. If one assumes, as is natural, that the E_n are energies (or else any monotonic functions of the energies), one readily arrives at a contradiction with experience which, to Hilbert, seemed fatal for any such scheme: From the preceding theorem one concludes that there cannot occur convergence of a series of spectral lines to a point within the observable spectrum, and this is in flagrant contradiction with experience where such points are observed. Only some years later Ritz discovered the fact that the frequencies of spectral lines and their corresponding energies are not proportional to the energy levels of atoms or molecules but are proportional to the *differences* between energy levels. This at once removes Hilbert's paradox—but by that time, Hilbert remarked, he had long since become preoccupied with other problems and had forgotten all about spectroscopy.

At the time of this incident, Schrödinger's series of basic papers had just begun to appear, laying the foundation for relating spectral lines to the theory of linear operators. As this work followed upon the heels of Heisenberg's great discovery of matrix mechanics (although the two were quite independently conceived) the ground had been well prepared for a very rapid acceptance of quantum theory. Similarly, the other major novelty of quantum mechanics, its statistical interpretation, would not have been adopted so readily by physicists had it not been preceded by the development of the powerful methods and insights of statistical mechanics a generation or two earlier. It was in particular the abstract and often purely algebraic manipulation of statistical propositions by Gibbs that made it easier to achieve the transition to an intrinsically statistical theory. The third major theoretical ingredient of quantum mechanics, namely, the Hamiltonian form of mechanics, had been part of the apparatus of physics for a long time already.

Thus we have the spectacle of an entirely new branch of science erupting, as it were, to the superficial beholder perhaps from nowhere, but on a more detailed view after the most intricate preparations. There were also guideposts concerning the use of this novel tool. One is safe in saying that *all* energy levels can be calculated as solutions of certain eigenvalue problems, and all spectral lines or, more generally speaking, radiative transitions in whatever part of the spectrum can then be found as differences of these energy levels. The same argument that applies to radiation also holds for other atomistic entities, electrons or protons, involved in a single interaction with any atomic or molecular system. On a little reflection it becomes clear that, given the precise mathematical formalism, one can proceed a long way toward the quantitative picture of many atomistic phenomena using only a very simple, horse-and-buggy conceptual scheme of interpretation.

Still, it might appear on further consideration that quantum mechanics is not yet really integrated into the prevailing schemes of conceptual thought. Instead, we have something like a physico-mathematical technology that, to some extent, has gone out of control. Everybody knows that the "going out of control" is what has happened to so much of contemporary technology, to the extent that this fact is among the chief characteristics of the present age. The intellectual phenomena are not here thought of as being "caused" by economic events but are merely claimed to run parallel to them due to circumstances which can be studied as part of the history of science but which are certainly not meant as explanations in the sense in which this term is used in physical science.

The dissociation of science from its conceptual background is largely a result of this overly rapid development, and one of the ways by which we can hope to restore some equilibrium is through paying attention to the more conceptual—and thereby often more philosophical—aspects of physical science. So far as the importance of the conceptual elements in quantum mechanics in particular is concerned we can, in part at least, agree with the well-known physicist L. Rosenfeld when he says: "The ordinary language is *inseparably* united, in a good theory, with whatever mathematical apparatus is necessary to deal with the quantitative aspects. It is only too true that, isolated from their physical context, the mathematical equations are meaningless."[1]

Even though this sentence has a great deal of truth in it one can greatly improve on its clarity. First, to say that equations apart from their physical context are meaningless is true only if we restrict meaning to the interpretation of observations; but formulas do of course have a well-defined meaning insofar as they are considered as parts of a mathematical structure, the meaning being then merely one of tautology relative to the given axioms (but meaning nevertheless, in a broader sense). Second, the statement that ordinary language and mathematics must be "inseparably" united with each other (the italics in the quotation being Rosenfeld's) can only be conceived of as weakening what we may call *Mach's program.* As anyone acquainted with the history of modern physical theory knows, Ernst Mach was the first to establish the plan to eliminate systematically from scientific discourse all components that pertain to metaphysics. Many other men have followed in Mach's footsteps; still, given the historical situation, the term "Mach's program" seems appropriate. One readily perceives that the problems that arose in what is called "the theory of measurement" derive precisely from the type of difficulties that Mach's program was designed to avoid. We may conclude that the application of Mach's program to quantum theory is still far from adequate.

One should remember here that concepts are intrinsically loose; they cannot of course be defined with complete rigor as sets of abstract symbols that are relative to given axioms can be. The so-called "definitions" of traditional philosophy represent concatenations of terms whose meaning and syntax are sufficiently familiar to generate a feeling of satisfaction in the reader or hearer so that it fails to induce him to ask further questions. Now a scientific statement that contains reference to experience will always have some verbal components (in addition to whatever mathematics there may be) and therefore the question of whether there are any metaphysical elements in the propositions cannot even be answered in a categorical manner (quite apart from how we wish to define "metaphysics"). It is therefore meaningless, in a "rigorous" sense, to say that Mach's program has been carried to completion. Nevertheless, one can assert confidently that the application of Mach's program to quantum mechanics has in the past been inadequate. As I have shown elsewhere[2–4] a quite systematic application of this program leads one to consider quantum theory as a special form of the theory of inductive probabilities. There are two distinctive specializations: the appearance of complex phases of the "waves" from which probabilities are computed (a well-known feature of conventional quantum theory) and the requirement, in statistical analysis, that all classes of empirical objects to which the theory refers have finite membership. This last, somewhat intricate requirement has been extensively discussed in my writings, and I do not plan to take it up in this brief paper.

In the sequel I shall be concerned with the limited task of pointing out and so far as possible elucidating two conceptual fallacies that, owing to the overly rapid development of quantum mechanics, still remain in many places whence one should be able to dislodge them. One of these is the "subjectivist" fallacy which tries to carry the concept of a subject or the concept of consciousness into physical theory; the other is the "symbolistic" fallacy in which mathematical symbols used for description are invested with properties of physical objects beyond permissible limits. These notions will be explained as we go on.

The subjectivist fallacy has taken on various forms in the course of history. An old one is connected with the gradual and rather laborious process whereby physics has differentiated itself in a specific way from metaphysics. A onetime famous example of a scientist who chose to disregard this process is offered by the brilliant German physiologist Dubois-Reymond, who, around the middle of the last century, in his public lectures raised a highly sonorous issue about the fact that science cannot and never will be able to "explain" consciousness. In this, he had great public success. We see here a dramatization of the fact that science is not meant to serve as a goal for the profound and presumably inborn human urge toward the absolute, for the "Ground" of things. Science, of course, sublimates this drive and thereby makes it useful, but science cannot "satisfy" such drives in an immediate manner—it was never meant to do this.

We are here only concerned with that part of philosophy which relates directly to physics, and this is of course epistemology, the theory of knowledge. When it comes to the relations of science and philosophy there are almost as many points of view to be found as there are writers on the subject, and so we might be permitted to present one too.

If philosophy were merely a sort of metascience, a science of sciences, as some conceive it, one could well do without it. The function of science itself has been well enough defined by those who have studied this question: Science deals with the *ordering of observations* by means of more general propositions, and from these the outcome of special cases may often be deductively derived. There is also a purely descriptive side to science which is usually (and in the main unfortunately) given a rather inferior position. Now if the truth of a proposition is derived from its comparison with observational data, it belongs by definition into one of the empirical sciences; obviously philosophy cannot merely be concerned with truth on an empirical basis; what then is it concerned with short of purely speculative metaphysics? I have slowly become convinced that by far the best way of stating the function of philosophy in relation to science is in *biological* terms: *Philosophy in its relation to science subserves an adaptive function.* Since science clearly pertains far more to society than it does to the individual, this implies that the philosophy of science has the function of adapting scientific facts and results to use in the context of evolving society. This does not at all imply that philosophy should concern itself with the practical uses of science. Since science is generally speaking an intellectual instrumentality it needs to be understood and interpreted as such within the context of intellectual life of society. We have thus nothing at all in common with those who, in a more medieval fashion, look for a "normative" function of philosophy. In this regard, we have been greatly impressed by the work of the biologist W. H. Thorpe[5] who thinks that biology will in the future play the central role relative to the sciences which metaphysics played in the past. (Thorpe says "theology" where we have substituted metaphysics without, we hope, violating his meaning.)

In the philosophy, i.e. epistemology, of science we are primarily indebted to those scientists who have made the effort of delving deeply into such epistemology and who have come up with significant results that help often to combat entrenched misconceptions. These men include, first, the Viennese school, Ernst Mach, and later Philipp Frank, and others. Later on arose the Cambridge school, Whitehead and his followers, Russell, Jeans, and Eddington. These investigators have reiterated the old truth that science is not concerned with qualities, which pertain to metaphysics, but *only with abstract relationships*. (The scholastics who were the first to clarify such ideas spoke of abstract relationships as "number and order," an expression which seems still altogether representative.) As Eddington[6] has put it with particular clarity, science is not concerned with consciousness at all since consciousness is a quality and science cannot deal with qualities; but if we find within the realm of consciousness regularities, i.e. abstract *structural order*, then this, and only this, is an object for science (see also Elsasser[7]). Furthermore, the British school has gone much beyond their predecessors in recognizing a point of major significance which may be set out as follows.

Science is not so much concerned with determining numbers and order for individual sets of observations as with building up large, consistent abstract structures which then, *as wholes*, will be compared with large bodies of observations. Basically, this approach is quite old: the first representative of a purely abstract structure is Euclidean geometry which is a branch of pure mathematics as well as an observationally verifiable empirical science (the latter fact being too often forgotten). But, historically speaking, it was only in rather recent times that mathematicians have brought to light a sufficiently rich array of abstract structures (groups, algebras, and other algebraic structures; the innumerable structures of set theory) so that such a program, which would be the final elaboration of what we have called Mach's program, makes perfectly good sense. Passing over most of the history of physics, we might say that quantum mechanics in its more abstract forms (as presented, for instance, by Dirac) does at the least approach this kind of structure.

In comparing such an abstract structure with bodies of observations, we find ourselves in the following position: Given a specific abstract structure such that, on empirical evidence, we may consider it as a systematization (theory) representing a large body of observations, the process of *deduction* from basic axioms becomes a set of clearly defined and, so far as the pure mathematics goes, unambiguous formal operations. The comparison with observation does then not, as a rule, engender any particular difficulties. The role of *induction* can be made very clear within such a program: it consists in the suitable *choice* of the abstract structure or structures, among a manifold which may be available to describe the body of observations at hand. While induction uses mathematical (probabilistic) techniques, we seem authorized, on the basis of the amplest evidence, to say that the inductive process cannot be fully formalized. It has been said often enough that science is always a mixture of inductive and deductive processes. The idea of a science from which induction has been fully eliminated, thus making science rigorous and infallible, is just a chimera.

Let us now turn back to the conceptual fallacies related to the interpretation of quantum mechanics. Of particular interest is the well-known interpretation of the quantum-mechanical process of measurement in terms of a partly irrational subject-object relationship, first proposed by Heisenberg and, for a while at least, accepted by many physicists. On a previous occasion[8] I have pointed out that it would be false to think of this as a basic novelty; it is instead the application of an already widely accepted philosophical mode of thought, the philosophy of Kant, to a special situation in physics. We find it, therefore, useful to glance at the relationship of Kantian philosophy to science rather than to try to understand the specialized form which this philosophy was given in its application to quantum theory. The story begins naturally enough with Newton's *Principia*, where we are surprised to see Newton engaged in unsubstantiated grand generalizations; he says (according to Chandrasekhar): "Every particle in the universe attracts every other particle with a force directly as the product of their masses and inversely as the square of their distance."[9] Astronomers among Newton's contemporaries and his followers felt very uneasy about this because the evidence at that time was wholly confined to the solar system; observational data from more remote stars became available only very much later, in the nineteenth century. If the greatest of all physicists was so prone to extreme generalizations one cannot wonder that we lesser people are equally so.

But a far more dangerous generalization concerns not the distant stars but the application of ideas of this type to the world of atoms. It is true that some men were extremely reserved and prudent in this respect, for instance Willard Gibbs, but a large segment of the scientific community never was. Earlier, in the eighteenth century in particular, philosophical minds had been greatly concerned with the consequences of a presumedly universal and completely deterministic causality of the Newtonian type. Such an idea is of course most uncomfortable for anybody except a confirmed, dogmatic mechanist. Leibniz in the philosophical writings of his older age was deeply concerned with overcoming mechanistic prejudices, using, not infrequently, what appears to us as somewhat artificial methods. Among the eighteenth century philosophers who tried to overcome the conflict between the speculative idea of universal determinism and the metaphysical concept of so-called "free will," one of the most outstanding was no doubt Kant. This is not the place to expound his philosophy; we can, however, say from the viewpoint of the scientist that Kant's work was extremely stimulating to scientific research in the generations that followed him. Kant's statement that epistemology is beyond the scope of the scientist's limited rationality and ought to be left to professional philosophers was quite direct; it no doubt encouraged a large number of scientists in the nineteenth century to investigate nature with little concern for philosophy, preferring cut-and-dried mechanical models. Of the eminent success of this approach *in that period* there can be no doubt. But to use this approach in the present-day, highly sophisticated condition of science is quite another matter: to employ the methods of the frontier when the frontier has vanished is a dangerous undertaking.

At this point the view of philosophy as the expression of a biological requirement for adaptation may be put to the pragmatic test. As mentioned before, if we deal with philosophy from the viewpoint of the scientist we are concerned mainly with epistemology, that part of philosophy which has a direct bearing on science. Of course, science itself is pragmatic, its propositions can be tested in terms of the outcome of experiments. The tests may become somewhat difficult when the propositions are essentially statistical, but it is possible even then to accumulate evidence for or against a given set of purely scientific statements. But philosophical propositions are not such that they can be demonstrated as factually true, nor can they be rejected on a simple pragmatic basis. We must hence try to conceive of the philosophy of science differently.

Whenever we deal with the study of an environment of some complexity, we cannot help but *project* the contents of our own mind upon the environment. Under more primitive circumstances this activity is not considered wrong but, in the contrary, a virtue. Thus the early Greeks saw a fairy queen presiding over every wooded grove and saw a dignified bearded gentleman sitting on the cloud-covered top of Mt. Olympus engaged in producing the phenomena of thunder and lightning. Far from criticizing this, we can take it as an expression of great imaginative powers. No intelligent person would question that we are dealing here with a form of projection, from the beholders into nature. But if in the course of development the population of Olympus becomes replaced by an assembly of Platonic ideas, universals, or simply general concepts, it is that much more difficult to see the projections, *precisely because we ourselves think to a large extent in the same categories, not because projections have simply vanished.*

One should not be tempted into thinking that the appearance of projection is artificially contrived or is a marginal affair at this point. Projection is one of the basic activities of the human mind; to express this alternatively and in more physiological terms, projection is one of the central functions of the human cerebral cortex. There is rather general consensus about the fact that the most distinctive function, and one of the most basic ones, of the human forebrain is symbol-making. Symbol-making is at the basis of language, communications, science, and practically all other collective human activities that involve higher mental faculties. But in the creation of symbols, sense impressions are not the only ingredients. The sense impressions, on being ordered by the mind, become inevitably suffused with subjective elements that, at their first appearance, are, as a quite general rule, unconscious. Projection is almost entirely an unconscious process; but not only that, it is one of the most important expressions if not the most important mental expression of unconscious drives. Carl Gustav Jung says, in fact, bluntly: "The unconscious *is* the projection." If a man of Jung's stature makes such an explicit statement, we can be reasonably sure that he knows what he is talking about. To come back to our initial remark that the philosophy of science deals with the relationship of science to other activities within the intellectual realm, this can now be reformulated: The philosophy of science tries to analyze the relationship of the symbolism of science to the broader aspects of the symbol-making activity of human society. Except in the case where symbols are purely abstract, i.e. logico-mathematical, and their meaning is directly verifiable in terms of concrete observations, i.e. when we deal with science in the most rigorous sense of the word, symbolic operations do virtually always contain projective elements. But since these projections begin by being unconscious and very frequently present resistances to being consciously analyzed, the brilliant intellects who work to create these symbolisms are often the last ones to recognize their unconscious contents. It is essential to appreciate the biological aspect of these intellectual tendencies, for if one characterizes them simply as "intellectual tendencies," one makes them appear quite pale and innocuous, and one ignores the extremely powerful biological urges toward symbolization and hence generalization that propel them. These are just as essential for the development and progress of the intellectual faculty as, say, the urge to play is essential for the development of muscular coordination.

Returning to the philosophy of Kant, we can now look at it in a different light. We see a quite arbitrary construct, that of universalized Newtonian determinism set up dogmatically as a "necessity of thought" and thus brought into conflict with a traditional but scientifically unverifiable concept, that of "free will." We can now see in this a projection of unconscious conflicts existing within the personality of the philosopher. By the available evidence, Kant must have been a quite complex and difficult person. On projecting one's own conflicts into grandiose generalities one does not resolve these conflicts but one makes it easier to live with them. On projecting in turn Kantian philosophy upon quantum theory one does not advance physics, but one may well mitigate some major psychological conflicts. The wide acceptance which Kant's philosophy had found, in central Europe in particular, can therefore be understood by saying that Kant succeeded in producing a sympathetic resonance in a large number of individuals whose unconscious conflicts were similar to his but who were far less capable of projecting these conflicts into generalizations. It stands to reason that similar arguments are applicable to many other philosophies. From the viewpoint of this analysis it would be erroneous to declare philosophical efforts that precede those of the present day as "antiquated." Philosophy does not in general deal with the external world in a way that can be conclusively verified by concrete operations and where therefore projection could be largely eliminated. Philosophies often project tensions existing in an individual or in a social body upon general conceptual structures, and in doing so they assist very greatly in making these tensions less severe. Reduction of tensions by related organic functions is an altogether basic biological process. Its well-known importance has recently again been stressed for the layman by the famous biologist, Konrad Lorenz[10] in connection with another drive which, while perhaps biologically older, is in man by no means more widespread than the symbol-making activity, namely aggression.

We should emphasize here that there is one form of intellectual activity that, in a sense, straddles science and philosophy and is rather more exempt from the danger of containing too much of unconscious projection. This is *inductive inference.*[11] It is essentially a method; so far as one can see, it cannot be made into a literal scientific "theory" with axioms and deductions. (Induction, moreover, fails even to approach scientific method if it is a form of generalization based on too small a number of given samples. If we generalize from a very limited set of samples, subjective errors and projections are bound to creep in and we are moving away from science. However, this subject is too broad to be dealt with here; it would require a much more detailed investigation.) In relation to induction we should note that Niels Bohr's frequent remarks on the ultimately macroscopic nature of all measurements are bound to play an important role whenever these arguments are applied to quantum theory. If a measurement is truly valid only if it leaves an impression in the *macroscopic* world (which latter, by consensus, is generally accessible, public, and in this sense "objective")—then one may conversely infer that microphysics can only be satisfactorily established by means of inductive inferences that proceed from the macroscopic toward the microscopic. It seems to us that this broad (and apparently incontrovertible) principle has not nearly received enough attention among physicists.

On continuing the study of logical fallacies in the interpretation of quantum mechanics, we shall next review the second type, namely, the "symbolistic" ones. As opposed to outright mechanistic prejudices which are usually easy to recognize, they are more intricate and arise out of false identifications of mathematical symbols with objects to which the symbols are related. Within classical physics, symbols may quite legitimately be taken as representing objects, and indeed often representing all available knowledge about the objects. But in quantum mechanics the relationship is not so simple because the most common symbol, a wave function, *does not refer to an individual object at all*; it describes the statistical distribution of a large number of equivalent objects, a *class*, and contains very little information about an individual object considered by itself, apart from the class. The truth is, as quantum mechanics has shown, that nature does not choose to provide us with tools for describing individual objects on the atomic level; we are only given tools to describe classes. To quote here Rosenfeld again: "The wholeness of quantal processes necessitates a revision of the concept of phenomenon. The concepts which in classical theory describe the state . . . can no longer be regarded as denoting *attributes* of the system . . . rather, they express *relations* between the system and certain apparatus of entirely classical (i.e. directly controllable) character. . . . It is perfectly true that this novel experience has profound epistemological implications."[12]

Here, Rosenfeld emphasizes the impossibility of describing an individual atomic object in a quantitative manner, whereas in the sentences just preceding we spoke of the possibility, which remains, of describing statistically the behavior of a class. The two things are, however, most closely related in a manner explained in every good textbook, therefore not in need of recapitulation. But this situation is awkward because it runs so much counter to our most ordinary mental habits. For instance, I am a rather pronouncedly visual type, and when I close my eyes and say the words "hydrogen atom," I can see before me a coordinate system and in it the familiar exponential which represents the probability distribution in the hydrogen atom's ground state. It takes a major effort of will to reassure myself that this exponential is not at all "the shape of *an* atom" but that it can be verified *only* by means of a lengthy sequence of atoms which will be destroyed in the process of measuring the position of their electron. Conceptual quandaries of this and related types are not confined to quantum mechanics, of course. They have been termed "intuitional paradoxes" by this writer;[13] they will appear every time a radical piece of scientific progress requires a reorientation of our habitual modes of thought.

The symbolistic transfer of quantitative meaning from the class to the individual has, however, specific mathematical implications, as was indicated particularly by E. Wigner.[14] He was considering the replication of biological macromolecules, and he succeeded in showing this: Given a system of a quite arbitrary constitution but containing one subsystem ("molecule") that is well defined and of some complexity. Assume that one requires that after some time this system has changed into one that contains two exactly identical molecules of the type of which there was originally only one. Letting both the initial and the final state be represented by a *single* wave function, Wigner shows that the transformation equations that lead from the initial to the final wave function, and that are uniquely given by quantum mechanics, cannot be satisfied, because there are as a rule far more equations than there are variables. This is of course an impossible result since the exact replication of molecules, of simple as well as of complex ones, takes place all the time in chemical factories. Wigner therefore suggests a revision of quantum mechanics. The underlying difficulty has been exhibited by Landsberg,[15] and it appears that the case belongs to the category of symbolistic fallacies, that is, a deficiency of conceptual manipulation in the relationship to data rather than to a failure of the algorithm. The problem does border on epistemology; we can here elucidate it by saying that a certain *duality is inevitable in any statistical description*, thus: If it is true that a wave function does not represent a proposition about a single atomic object but only about a class of equivalent objects then, conversely, one cannot expect that a single object be represented by one well-defined wave function; it can equally well be described by any of a large (usually infinite) set of wave functions which are compatible with the conditions of interaction under which the particular object is observed. The members of this set of wave functions will not differ from each other in the description of the basic structure of the molecule proper; they will differ in regard to subsidiary phases, technically "noise," which of necessity appears when the molecular object is observed by means of an (intrinsically macroscopic) measuring instrument.

We are thus confronted with what may be described as a general statistical duality: In a statistical theory, a symbol, in this case a wave function, does not describe a single object but only a class of equivalent objects; conversely, a single object cannot be described by a single wave function but only by a bundle of these, a statistical ensemble. This conclusion could only be avoided if we made the quite unrealistic assumption that the object is completely isolated and does not interact with other objects; but whenever the object does so interact, in particular when it is observed by a measuring instrument, a statistical ensemble of wave functions furnishing a description can always be found, not just one wave function. This by no means trivial fact points the way to an avoidance of symbolistic fallacies.

Fortunately, the mathematical technique of using ensembles of wave functions, that is, quantum statistical mechanics, is very highly developed. On applying it judiciously, epistemological fallacies, of which some have been enumerated above, can fairly safely be avoided. This should not lead to the idea that since the basic structure of the mathematical apparatus remains unchanged on such an analysis, the epistemological aspects are merely decorative, a frosting on the cake as it were. Errors of interpretation often remain irrelevant so long as the theory in question is only applied to a specifically limited field where a student can learn the rules and then produce satisfactory answers. But a good theory, more often than not, also serves as the basis of evolutionary progress, that is, as the starting point of quite novel theories; and to serve this function the theory must be adaptable. The literature of theoretical physics often contains papers in which a slight "modification" of some basic mathematical formalism is proposed, but in a well-rounded theory one cannot as a rule change one structural element while leaving all others unchanged, and hope for success. In physical science, the tool of adaptation is, more often than not, a thorough epistemological analysis. This is as true in quantum mechanics as it was once upon a time in classical physics (Mach) where the attendant successes are familiar.

But while in relativity there is no epistemological problem about symbols representing "real" objects (although certain sets of symbols are capable of transformations described as "covariance") the abstract structure of quantum mechanics is altogether different, since the theory deals with statistical relationships. The question will then arise: What are the epistemological problems in a statistical system of description? This writer has spent many years trying to analyze these problems. Ultimately, the pivotal questions turned out to be two in number: first, the relationship between individual objects and abstract classes designed to represent properties of objects in the theory; second, the size of such classes, that is, the number of their members. The size of a class can of course range from having only one member ("individuality") over having several or many members (but a finite number) to the mathematically convenient case of infinite membership. My principal conclusion is that the epistemological difficulties which Mach's program is expected to elucidate and ultimately to eliminate, are buried from view by the use of classes of infinite membership. In accordance with the idea of astronomers that the universe is finite, we must represent the world by abstract structures of finite size in which, then, all classes have finite membership. (This does not of course preclude the use of infinite classes as purely mathematical tools in many cases so long as one exercises due caution in their physical interpretation.) This conclusion, which I called the "principle of finite classes" appeared in my first book (1958) and is repeated in all my later writings. Since at that time I wrote primarily for readers versed in the life sciences, I tried to avoid mathematical formalism, but more recently I have returned to a somewhat more formal mode of presentation. The so-called theory of inductive probabilities, developed systematically only in this century, turns out to be a perfect tool for quantum-statistical description. I have recently summarized this abstract development.[16] It is true that much more technical work will have to be done before such methods could be applied to specific biological situations. The only apparent exception, as I have pointed out earlier and again lately,[17] is the study of morphological and physiological individuality in higher organisms, a branch of biology greatly undervalued in the past. Still, we need not wait for detailed biological applications to draw from these abstract principles significant philosophical conclusions as I have indicated recently.[18] It then appears that the deductions drawn from these highly abstract and "scientific" modes of reasoning have some similarity with ideas that now, as it were, float about in contemporary society. For those whose philosophical thought runs along lines similar to the ones sketched above, this will not be too surprising.

We have recently heard again from an old champion of these intellectual battles, none other than Albert Einstein.[19] In a number of his letters to Max Born but perhaps most emphatically in one dated December 3, 1953, a year before his death, Einstein has this to say: If his (Einstein's) old desire to revert to classical determinism cannot be fulfilled, then one better interpret quantum mechanics as it demands to be interpreted and that is, in Einstein's view, in terms of *ensembles*. This means that nature in the small does not exhibit a definite, numerically describable pattern; it does so only in the macroscopic, classical limit. If I dare modify the terminology, in this view the laws of nature are not descriptions of an objective reality; they represent *constraints* on a reality that is neither defined nor definable above and beyond these constraints.

References

1.
L. Rosenfeld, in *Observation and Interpretation*, ed. by S. Körner (Academic Press, 1957), p. 41.

2.
W. M. Elsasser, J. Theoret. Biol. **25**, 276–296 (1969).
3.
W. M. Elsasser, Am. Scientist **57**, 502–516 (1969).
4.
W. M. Elsasser, in *Towards a Theoretical Biology*, ed. by C. H. Waddington (Edinburgh University Press, 1970), vol. III, pp. 137–166.
5.
W. H. Thorpe, *Science, Man and Morals* (Cornell University Press, 1965).
6.
A. Eddington, *The Philosophy of Physical Science* (University of Michigan Press, paperback ed., 1958) (1st ed., 1939).
7.
W. M. Elsasser, *The Physical Foundation of Biology* (Pergamon Press, 1958), chap. 5.
8.
W. M. Elsasser, *Atom and Organism* (Princeton University Press, 1966).
9.
S. Chandrasekhar, Notes of a Public Lecture, University of Chicago, 1966.
10.
K. Lorenz, *On Aggression* (Bantam, paperback ed., 1965).
11.
Ref. 2.
12.
Ref. 1.
13.
Ref. 8.
14.
E. P. Wigner, *The Logic of Personal Knowledge*, Essays presented to M. Polanyi (Routledge and Kegan Paul, 1961).
15.
P. T. Landsberg, Nature **203**, 928–930 (1964).
16.
Ref. 2.
17.
Ref. 4.
18.
Ref. 3.
19.
Albert Einstein, Hedwig und Max Born, *Briefwechsel 1916–1955* (Nymphenburger Verlagshandlung, 1969).

Fifteen **Léon Rosenfeld**

Unphilosophical Considerations on Causality in Physics

1. Causality in Classical Physics

Historically, the elaboration of the scientific concept of causality as a fundamental element of rational thinking is a by-product of the development of the formulation of the laws of motion of material bodies and their successful application to the dynamics of the solar system. Especially this last circumstance strongly influenced early epistemological reflection in encouraging radical idealizations, such as the law of inertia and the resulting notion of force as producing acceleration and thereby uniquely determining the motion. Thus, the concept of physical causality became endowed right from the start with the connotations of necessity and determinism. The belief in the adequacy of this deterministic causality was strengthened by the initial success of the attempt to reduce all physical phenomena to mechanical processes; such a view of the physical world was not necessarily atomistic, but it accepted the basic idea that all forces were essentially contact interactions between elements of matter. It is true that the force of gravitation, and later those of electricity and magnetism, were treated as long range interactions. This, however, was regarded as a phenomenological description, which ought to be reduced to some local form of interaction between the elements of ordinary matter and those of some subtle medium through which such forces could be propagated even in the absence of gross matter. Although this radically mechanistic view had to be abandoned, and the transmission of electromagnetic and gravitational interactions ascribed to autonomous agents, the basic causal structure of physical theory in its final classical form was fully upheld in this theory: all interactions of the material atoms and atomic constituents with the fields of electromagnetic and gravitational force are strictly local and lead to causal relations of a deterministic character.

Let us at once emphasize how remote this abstract form of physical causality is from the immediate perception of causal relations manifested by adult subjects when they are confronted, as in the classical experiments of Michotte and his school,[1,2] with moving visual patterns simulating the collision of solid bodies. Not only is a causal relationship perceived even when the collision represented is dynamically impossible, but this relation is felt to arise before the contact and subsist after it within finite intervals of separation of the colliding bodies—their "action radii"—depending on their velocities. When the body to which momentum has been transferred has moved away from the other farther than its action radius, its motion is perceived as autonomous and no longer related to the collision (unless the velocity it acquires is considerably larger than the initial velocity of the other body). Indeed, the descriptions of the process perceived, given spontaneously in common language by the subjects, are strikingly reminiscent of an ancient and medieval conception of the causality of motion, of which we find the earliest historical evidence in the work of the Byzantine scholar, Johannes Philoponos.[3] Just as the latter argued that the motion of a stone or an arrow is due to an incorporeal "kinetic power" (later called "impetus") communicated to the mobile by the hand or the bow, the subject speaks of "motion" as of something with which a moving body is endowed and which may pass by contact onto another body. The finite action radius of perceptive causality corresponds to the view expressed by Philoponos that the supply of kinetic power given to the mobile in the initial contact with the mover is finite and can only maintain the motion over a finite distance.

A further feature common to perceptive causality of the transmission of motion and to the medieval conception is the complete absence of any notion of inertia: the transmission of motion is conceived as a one-sided process, in which one body acts upon another to set it in motion, or modify its motion; the latter body thereby remaining entirely passive and without influence on the process. This last peculiarity of the type of perceptive causality exhibited by adults is the more remarkable as it is in regression from the causal anticipations and explanations of mechanical processes to which children are led in the course of their spontaneous mental development. In fact, the investigation of the formation and elaboration of schemes of causal relations by children, currently pursued by Piaget and his school,* shows that children do take account of the fact that a body exerts more or less "resistance" to being set in motion—an attitude quite in harmony with the general predominance, in the child's thinking, of experience derived from his activity over purely sensory elements. A study by R. Droz,† in particular, reveals that this notion of resistance of a body to being moved, which may be regarded as a prefiguration of the Newtonian concept of inertia, is even incorporated by some children in their spontaneous description of the perception of Michotte visual patterns simulating transmission of motion. It would seem, however, that this link between the perception of causal relations and an improved dynamical representation of the processes does not persist, and that the adult reverts to a more primitive form of perceptive causality, dominated by the sensory pattern.

An essential feature of this pattern, emphasized by Michotte, is its global character; as soon as a break occurs in the continuity of the process represented, any perception of causal relation disappears. This global character of perceptive causality is the opposite of the analytical mode of description of classical mechanics, in which the motion is indefinitely subdivided into elements characterized by instantaneous positions and momenta of the moving bodies. Here the relation of causality expresses the existence of a necessary and unique dynamical link between neighboring

* These problems were recently discussed at the thirteenth symposium on genetic epistemology held in Geneva in June 1968. I take this opportunity of expressing my gratitude to Professor Piaget for enabling me to attend this symposium and report on the epistemological situation in modern physics. The present paper is a version of this report, revised in the light of the new information discussed at the symposium.

† This study was presented at the thirteenth symposium on genetic epistemology mentioned in the preceding footnote.

elements of the motion. Necessity and uniqueness are foreign to perceptive causality: a slight variation of the circumstances suffices to wipe out the perception of a causal link in the global pattern observed. There is no question, therefore, of representing (as some physicists uncritically do) the abstract form of deterministic causality of classical physics as corresponding to some deep-lying requirement of our mental setup. Certainly, the concepts of classical physics may be said in a loose way to originate from "common experience"; at any rate, the systematic experimental studies of child behavior initiated by the Piaget school have already traced some of them, in a rudimentary form, to their origin in the sensorimotor schemes built up by the child in the course of his mental development; but their final elaboration demands the highest power of abstraction of which human minds are capable. They represent idealizations exquisitely adapted to a symbolic, codified description of the domain of experience ultimately accessible to us through sensory perception; but such a refined code system is not the one we need for the guidance of our day-to-day behavior in the physical world: for this purpose, unsophisticated sensorimotor schemes still suffice in our scientific age, as they sufficed for the millennia of human history preceding the emergence of modern science.

In a sense, Lagrange was right to describe "analytical mechanics" as a branch of mathematics; but he made a fatal error of judgment—shrewdly pointed out by his contemporary Poinsot[4]—in ascribing to mathematical reasoning the power of establishing the necessary truth of physical laws of absolute generality. He underestimated the danger inherent in the idealizations that make the mathematical formulation of such laws possible and give them such sharp precision and supreme elegance: this result is only achieved by fixing the scope of the idealizations according to mathematical criteria, at the risk of extending it beyond the limit at which these idealizations cease to give an adequate representation of the physical phenomena they purport to describe. Increased experimental knowledge of the phenomena may thus force us to assign limits to the validity of a physical concept; such a step is not only a progress in scientific theory at the practical level, but also a major epistemological advance, since no physical concept is sufficiently defined without the knowledge of its domain of validity. A case in point is the law of inertia itself, whose simple Newtonian enunciation, far from being universal, is now recognized as only the asymptotic form of a more general one, which is part of a theory taking into account physical conditions for spatiotemporal localization, ignored in Newton's mechanics. From this point of view, the deterministic type of causality of classical physics also appears as an idealization rooted—not (as we have seen) in any psychical experience, still less in any innate property of the mind (that will-o'-the-wisp of philosophers)—but in highly sophisticated experience of the physical phenomena. Accordingly, the determination of its domain of validity is not a matter of arbitrary decision, but entirely one of inference from experiment.

2. Conservation Laws

So far, we have discussed the causal relations of classical physics as they appear when one looks at the evolution of the phenomena in space and time. There is, however, another aspect of causality which is not concerned with spatiotemporal localization of a physical process, but with the balance of energy and momentum exchanged between different parts of the system in the course of the process. In its full generality, including the phenomena of heat and electromagnetic radiation as well as purely mechanical processes, this accountant's view of causality is surprisingly modern. During the middle ages and the Renaissance, the dream of a *perpetuum mobile* was entertained even by the most skillful and experienced craftsmen—like that thirteenth-century church builder, Villard de Honnecourt, who jotted down the sketch of such a contraption in his notebook.[5] When it was banished from the realm of mechanics in the eighteenth century, the secret hope of discovering some inexhaustible source of power persisted in less well-known domains of physics. Thus Joule himself spent much effort in constructing electromagnets because he originally expected that, with increasing intensity of the exciting current, the power he could gain from the magnet would increase more rapidly than that needed by the battery to produce the current.[6]

Even in the domain of purely mechanical phenomena, the long and confused controversy during the eighteenth century about the conservation of "motion" in collision processes had not thrown sufficient light on the role of the forces operating the transmission of motion between interacting bodies, and the existence of a common measure for force and motion, regulating the transmutation from the one into the other, remained quite unfamiliar down to the middle of the nineteenth century. Accordingly, the idea of an equivalence between heat and mechanical work was also foreign to the then prevailing mechanistic conception of the physical phenomena, even for those physicists who (like Laplace) were perfectly aware of the possible identification of heat with molecular motion. When the young engineer Colding submitted to Oersted his thoughts about the production of heat by friction, which he wanted to test by large-scale experiments, Oersted's reaction betrays a confusion characteristic of this general attitude: there can be no relation between the force applied and the heat produced, he objects, since, assuming heat to be due to some kind of motion, the total momentum of this motion (which he takes as its measure) is always zero; no applied force can be "lost"—this was d'Alembert's conception—except by being compensated by the reaction of some constraint. (Nevertheless, let it be said to his and his colleagues' credit, they did give Colding all the support and help he needed to perform and publish his experiments!)[7]

This unpreparedness of the mechanistic school for the aspect of the physical processes expressed by conservation laws is very instructive: it convincingly shows that this aspect was quite a novel one, of which the most experienced masters of rational mechanics were not conscious. Of course, once attention was focused onto these conservation laws, it was easy enough to exhibit them as logical consequences of the laws of motion, and to define the concepts of kinetic and potential energy in order to describe the transmutation of energy continually going on between work and motion in any dynamical process; the fact remains, however, that the existence of this transmutation had to be pointed out to the chevroned physicists by young outsiders whose vision was not obscured by too much knowledge. From the epistemological point of view, this means, obviously, that in spite of the trivial (and, as we shall soon see, misleading) circumstance that they are formally interrelated, the two aspects of the causality of physical processes arising from the consideration of their evolution in space and time, on the one hand, and of the exchanges of momentum and energy taking place in them, on the other, are actually independent of each other and correspond to two different ways of looking at the phenomena, both equally necessary for obtaining a complete picture of them.

3. Complementarity in Quantum Physics

Within the scope of classical physics, the logical relation between the two aspects of causality we have discussed does not seem to present any problem; and indeed, as soon as mechanistic physicists of the nineteenth century realized that the equivalence of work and heat was a simple consequence of the interpretation of heat as the kinetic energy of molecular motion, they were confident of having "reduced" the laws of heat phenomena to the Newtonian laws of motion. The extension of these laws to the atomic domain, where their applicability could not be directly tested, was not regarded as problematic, since the unlimited validity of the idealized concepts of mechanics was taken for granted. Mach's opposition to atomism, although (like all long-range predictions of prophets) it appears misguided to our increased wisdom, was actually based on a pointed questioning of this very extrapolation of our usual space-time description to atomic systems.[8,9] We are not justified, he argued, in attributing to these hypothetical constituents of matter, which we neither can see nor touch, a location or a displacement in space, since these notions only acquire a meaning by being referred to sensations of sight and touch, in which they originate. We should therefore not expect the parameters needed to specify the state of an atomic system to be the same as those describing a system of ordinary material bodies. The causality governing atomic systems need not be of the spatiotemporal type; it could have the more general form (of which the conservation law of energy is an example) of a set of relations between parameters of a different kind. This argument suffers from Mach's general tendency to envisage the relationship between physical concepts and sense impressions as more direct than it actually is. Anyhow, it was too inconclusive to make an impression on the advocates of a representation of the phenomena that had proved so successful. They went on in their own naïve way, taking their description of the molecular motions as an account of the spectacle that would offer itself to a human observer who, like Alice, would have eaten a bit of the right-hand side of the mushroom.

The causality problem took an unexpected turn, however, after the discovery of the quantum of action and the subsequent development of a theory of atomic constitution and atomic processes incorporating in a rational way the content of the quantum postulates and the correspondence principle. This theory operates with the same physical idealizations—space-time localization and momentum-energy exchange—as classical physics, for the simple reason that the conditions under which individual atomic processes are observed are essentially the same as those of ordinary observation. All the information we are able to obtain about atomic processes has ultimately to be referred to the indications recorded by experimental devices whose functioning is entirely described in classical terms. Now, the new feature introduced by the existence of the quantum of action is the occurrence of a mutual limitation in the use of the two modes of description just mentioned, which in classical physics, when quantal effects are ignored, become logically compatible without any restriction. It is important to realize that the limitation in question is not due to any imperfection of the classical idealizations: within its scope (excluding the domain of phenomena in which the structural properties of nucleons and mesons come into play), quantum theory is just as exquisitely adapted to the account of the experiments as classical theories are within their own scope. There is no limit to the separate application of either space-time localization or momentum-energy conservation to an atomic system, and there is no question of eliminating either of them, since both correspond to significant and distinct aspects of the system. However, the conditions of observation allowing us to localize the system and those allowing us to set up the momentum-energy balance of the processes taking place in it are no longer compatible—they are mutually exclusive.

We are here confronted with a peculiar epistemological situation, which is not of our own choice, but is imposed upon us as a straightforward consequence of the existence of the quantum of action. In order to remind us of its implications, the concept of "complementarity" has been introduced: it denotes a relation between two physical phenomena, both exhibiting aspects of a given system indispensable for a complete account of its behavior, but whose conditions of observation are mutually exclusive. It is necessary, in view of the occurrence of this new type of relation, to include in the definition of a "phenomenon" the complete specification of the experimental arrangement by means of which it is observed; an essential item of such an arrangement is a device in which the characteristics of the process are registered in some codified form, thus ensuring the complete objectivity of its description. It cannot be too strongly emphasized that the concept of complementarity (like all our concepts) is just an element of a convenient code system, serving to warn us of the caution to be exercised in the use of classical physical concepts in order to avoid logical contradictions. Far from implying any deviation from ordinary logic, it helps us to keep our terminology in harmony with the strict logical deductions of the mathematical formalism. If one chooses to speak of "contradiction" in connection with the mutual exclusion of complementary phenomena, then one can only mean a "dialectical" contradiction; in fact, one may regard the occurrence of complementarity relations in quantum theory as a clear and precise example of what is implied by the general conception of "dialectical process." The quantum theory, which comprehends the complementary aspects in a logically consistent whole, represents the synthesis resolving the dialectical opposition between these aspects.

The two complementary modes of description of quantum phenomena correspond in classical physics, as we saw, to two independent types of causality, both deterministic. In order to achieve the synthesis of these complementary aspects, quantum theory abandons the excessive idealization of determinism and replaces it by the more comprehensive and flexible type of causal relations provided by the mathematical formalism of probability theory. From the logical point of view, this step is in harmony with the dialectical character of the theory; the logical formulation of a dialectical contradiction, according to Apostel's analysis,[10] essentially involves an element of probability. From the physical point of view, the use of a statistical mode of description is immediately suggested by our most common experience in the investigation of atomic processes. When we have fixed all the controllable circumstances of an experiment, we usually observe, not one definite type of process, but a great diversity of phenomena; under such conditions, we can expect to find regularities only in the statistical distribution of the various properties observed. Thus, it was evident since the earliest formulation of the quantum postulates that radiative transitions between stationary states of atomic systems could only be characterized by the rate of probability for their occurrence, since there is nothing in their definition that could determine the time at which such a transition would take place.*

* This was repeatedly mentioned by Niels Bohr in conversation.

Indeed, the concepts of stationary state and monochromatic radiation are extreme idealizations excluding every time determination, and the situation referred to in the quantum postulates is strictly complementary to the time evolution of radiation processes.

4. Structural Identity of Atomic Systems

The statistical method necessarily excludes from consideration individual differences between the objects with which it deals; it is concerned with the properties of species, not of individuals. It is this feature that most obviously distinguishes causal relations of the statistical type from classical causality. That a statistical method should be appropriate to the analysis of the properties of matter in bulk in terms of atomic processes, is merely a consequence of the fact that the conditions of observation of these properties exclude all possibility of determining the underlying dynamic behavior of the atoms. We have here two modes of description that are complementary, not on account of quantum limitations (which play no part in this argument), but because of the definitions of the quantities characterizing the macroscopic states of a physical system. In addition the formulation of the laws governing such states involves the neglect of all effects of the deviations of individual atoms from their average behavior under the given external conditions. Accordingly, the atomistic description of the phenomena (as was clearly recognized by Maxwell and Boltzmann) has to assign all the constituent atomic systems of a given species identical structures and to treat them as indistinguishable and interchangeable.[11,12]

In this idealization of identical atoms there is in the first instance an element of arbitrariness, inherent in the statistical method: How must we define the atomic or molecular species, whose members shall have identical structures? Obviously, this definition will depend on the experimental possibilities for establishing such specific distinctions. The history of chemical "philosophy" since Lavoisier displays a succession of advances in structural representations of molecules, each reminding us of the relativity of the concept of molecular species and of the pitfalls into which many a skillful chemist stumbled, under the illusion that he was sticking to "facts" and shunning "hypotheses." Thus, Stas believed that his accurate determinations of atomic weights disproved the hypothesis of a common building unit for the atoms of the chemical elements. Kekulé then showed a better grasp of the problem when he objected that the results of Stas were equally compatible with the hypothesis of a statistical distribution of the masses of the individual atoms of the same element. He added, furthermore, that it would then be conceivable that reactions could be found leading to a segregation of the heavier and lighter atoms—a pure speculation, he was anxious to stress, but it was a speculation based on sound logic.[13]

Atomic theory, however, has now reached the stage in which the conditions of observation of individual atomic processes have put an end to speculation and removed all arbitrariness. We may now regard as firmly established the identity of the constituents of atomic systems, electrons and nucleons, and the identity of structure of the atomic systems themselves, in all processes involving momentum and energy exchanges small enough not to affect the stability of the constituent particles. Indeed, in conjunction with the quantum laws to which atomic particles and systems are subjected, the identity of these elements acquires a fundamental significance, by leading to definite limitations of the possible symmetry properties which an ensemble of such interchangeable elements exhibits in its various quantal states. The specification of these symmetry properties is the key to the interpretation of types of interaction and correlation entirely outside the reach of classical ideas, and playing an essential role in the economy of nature: on these interactions and correlations depend, among others, the most important molecular bonds, the properties of the metallic state, as well as the striking phenomena of superconductivity and superfluidity appearing at low temperatures. The foundation of this vast body of knowledge would be wiped away if some "hidden parameter," embodying any further specification of atomic states than that given by quantum theory, crept out into the open.

The statistical causality of quantum theory refers to a deeper level of analysis, logically independent of the statistics of matter in bulk we have just been considering, and it affects the classical notion of a particle in a much more radical way. In the space-time representation of the motion of a particle, there is no place any longer for the classical picture of a continuous temporal succession of positions of the particle along a spatial trajectory; the link between successive observations of the spatial position of the particle is purely statistical. This not only eliminates from the quantal description of an atomic process any possibility of individualizing the particles taking part in it, but even of ascribing to these particles the permanence they would still retain in a classical description. Indeed, the process is completely characterized by indicating the numbers of particles of each species initially present in the various states they can occupy (which may, for instance, be spatially localized) and the changes these numbers undergo in the process. In the domain of atomic physics, these changes normally conserve the total number of particles of each species, but this is a much weaker statement than is implied by the classical notion of permanence of the particles. There is no controllable way in which the changes in the occupation of the states could be ascribed to transitions of definite particles from one state to another. A noncommittal way of accounting for the immediate data of observation would be to say that particles "appear" in certain states and "disappear" from others; one prefers to speak of "creations" and "annihilations" of particles in the states in question. The connotation of activity contained in this terminology has the merit of recalling that the observed appearances and disappearances are due to actual dynamical interactions. However much this quantal outlook departs from the classical picture of the evolution of systems of material bodies in space and time, it fulfills in the domain of atomic physics exactly the same function as the latter in its own domain: it is the only mode of objective conceptual representation of the phenomena that is accurately adapted to the experimental conditions.

The impossibility of attaching to the constituents of the atomic world the attribute of permanence inherent in the concept of material body used in classical physics will appear in its proper light when we realize how extreme an idealization this abstract notion of permanence involves. Here again, it will be instructive to turn to the psychology of perception; one aspect only of the very complex problem of perceptive permanence* need detain us. The typical situation exhibiting the perception of permanence of motion is the "screen effect": if the subject is shown a red square, along a side of which a white narrow band is made to appear and expand gradually in the direction perpendicular to this side, what the subject perceives is a solid white band emerging from behind the red square that hid it from view. A corresponding response is obtained with the inverse pattern, in which the white band initially present alongside the red square is made to contract and vanish: the subject perceives the band gliding solidly behind the screen. Now, one may combine the two patterns along two opposite sides of the red square, in such a way as to simulate the entry of the white band on one side and its exit on the other; one then obtains a remarkable "tunnel effect." If the motions of entry and exit are the same, and the time interval between them has the right order of magnitude, the subject perceives the motion of the band as permanently continued inside the tunnel; if the entry-exit time interval increases slightly, the impression of permanent motion persists—only, there is a "hitch" somewhere inside the tunnel, at a place which, curiously enough, the subject indicates with precision. However, this perception of the permanence of the motion disappears rather sharply when the entry-exit time interval reaches a critical value (depending on the other circumstances of the experiment). The subject then finds that the motion of entry "dies out" and that a new motion starts somewhere near the exit side,

* A thorough discussion of the whole problem, including the recent investigations of the Michotte school, will be found in the works by A. Michotte and his collaborators, cited above.

with an "empty space" in between. Graphical representations of the perceived motions, drawn spontaneously by some subjects, show in the former case a continuous trajectory, in the latter two branches of trajectory with an intermediate gap. Continuity of the motion, therefore—as a direct or indirect* component of the sensory pattern—is an essential condition for ascribing permanence to it in ordinary perception. If this element of continuity is not present under the conditions of observation, the unsophisticated observer reacts in much the same way as the atomic physicist.

5. By Way of Conclusion

The assessment of the epistemological novelties of quantum theory is often obscured by the belief that the causal structure of classical physics is such a compelling exigency of our *Anschauung* that it is the mold in which every physical theory should necessarily be cast. The experimental study of perception clearly shows that this belief is unfounded: the needs of ordinary perception of causality are satisfied by a mental representation of the motion of bodies much more primitive than the abstract conceptual framework of classical mechanics, which in fact has to respond to the exacting requirements of a rather singular mode of interaction of man with his environment. Nevertheless, the two responses have a common origin, and it is important to understand how such a sharp differentiation between them could arise, lest one draw unwarranted conclusions from its presence.

The decisive point is that mental development does not proceed along one uninterrupted line, but by cycles or stages, each of which

* This refers to the distinction proposed by Michotte and Burke between the "modal" and "amodal" data of sensation: the former are part of the immediate sensory content, the latter are only derived from it, as for instance the invisible part of the motion in the screen and tunnel effects. Piaget's genetic investigations indicate that such amodal elements may arise from the coordination of sensorimotor schemes in the child's early mental development: the screen effect is manifest in the behavior of ten-month-old infants.

leads to temporary harmony between the child's limited environment and the system of mental operations enabling him to cope with it. The ultimate stage is that ending with the acquisition of the "formal operations" which remain the tools of adult logical thinking. The attainment of this stage is prepared by the preceding one, the stage of "concrete operations," based on the direct use of the sensorimotor schemes of perception.

The characteristic difference between the two stages lies in the function of language. At the stage of concrete operations, language is simply a means of communication of sensorimotor experience: words are incorporated into the significant combinations of sensorimotor schemes, and constitute a symbolic representation of the latter. Since the sensorimotor schemes are themselves symbols of the aspects of experience retained as significant, one may say without excessive schematization that two correlated sets of symbols, or code systems, are used for the registration and communication of experience.[14,15] At the formal stage, however, the verbal code system pursues an autonomous development by purely abstract derivations of new concepts without immediate correspondence in the sensorimotor field. It is this formal refinement that makes scientific thinking possible, while it is obviously of little use in daily perceptive experience: hence, in particular, the discrepancy we noted between perceptive and physical causality. This detachment of the formal code system from the concrete one does not proceed without raising its problems, above all the ever recurring one of the adequacy of conceptual constructions whose link with sensorimotor experience is only an indirect one. The only way to clarify such questions is to go back to those basic idealizations which remain firmly anchored in perception, and to examine to what extent they can be consistently related to the other formal elements of the theory. In this respect, classical and quantal theories are in the same situation and the arguments by which their respective domains of validity have been bounded follow exactly the same pattern. Only, the way from idealization to sensory experience is a bit more roundabout for an atom than for a planet.

References

1.
A. Michotte, *La perception de la causalité* (Publications Universitaires de Louvain, 1954).
2.
A. Michotte et al., *Causalité, permanence et réalité phénoménales* (Publications Universitaires de Louvain, 1962).
3.
E. J. Dijksterhuis, *Val en Worp* (Noordhoff, 1924), pp. 36–40.
4.
L. Poinsot, "Théorie générale de l'équilibre et du mouvement des systèmes," Journal de l'Ecole Polytechnique (1805); reprinted in L. Poinsot, *Eléments de statique*, 9e édition (Bachelier, 1848).
5.
Album de Villard de Honnecourt (Bibliothèque Nationale ms. fr. 19093, f°5r°); reproduced in P. Mérimée, *Etudes sur les arts du moyen-âge* (édition Flammarion, 1967), p. 237. See further L. Whyte, *Medieval Technology and Social Change* (Oxford University Press, 1962), pp. 129 ff. and related notes.
6.
J. P. Joule, *Scientific Papers* (Taylor and Francis, 1884), vol. 1, p. 14 (from a letter published in 1839).
7.
V. Marstrand, *Ingeniøren og Fysikeren Ludvig August Colding* (Danmarks Naturvidenskabelige Samfund, 1929), pp. 23–24.
8.
E. Mach, *Die Geschichte und die Wurzel des Satzes von der Erhaltung der Arbeit* (Calve, 1872).
9.
E. Mach, *Principien der Wärmelehre* (J. A. Barth, 1896).
10.
L. Apostel, "Logique et dialectique," in *Logique et connaissance scientifique*, ed. by J. Piaget, Encyclopédie de la Pléiade, vol. 22 (Gallimard, 1967).
11.
L. Rosenfeld, "On the Foundations of Statistical Thermodynamics," Acta Phys. Polon. **14**, 3 (1955).
12.
L. Rosenfeld, "Questions of Irreversibility and Ergodicity," Rend. Scuola Intern. Fis. "Enrico Fermi" **14**, 1 (1962).
13.
J. Gillis, "Auguste Kekulé et son oeuvre réalisée à Gand de 1858 à 1867," Acad. Roy. Belg., Classe Sci., Mém.: Collection in 8° **37**, fasc. 1 (1966).
14.
L. Rosenfeld, *The Method of Physics* (UNESCO report, 1968).
15.
I. P. Pavlov, "An Attempt to Understand the Symptoms of Hysteria Physiologically" (1932), in *Conditioned Reflexes and Psychiatry* (International Publishers, 1941), pp. 113–114.

Sixteen **Hermann Bondi**

Logical Foundations in Physics

The progress of science depends on a wide variety of factors influencing the mutual stimulation of scientists. The interaction of theory and observation, the accidental (and sometimes even much later unexplained) failure of an experiment, the effect of the "climate of opinion," the particular mathematical knowledge of an individual scientist, all contribute to what might be called the "micro-history" or the psychology of scientific development. As a consequence of such incidental features it may readily occur that a whole section of the subject becomes established and accepted as the result of a circuitous and difficult path. This matters little to the individual research worker within this field since his concern is with the deductions to be made from the accepted view and its experimental implications. Thus there is a widespread indifference to attempts to put accepted theories on better logical foundations and to clarify their experimental basis, an indifference occasionally amounting to hostility. In my view it is a great pity that this attitude has severely limited the number of scientists working on such lines, and indeed is harmful to science. In a narrow sense this harm arises because, owing to our ignorance of the experimental and logical foundations of theories, we are caught by surprise more than is necessary when the next advance shows up the limitations of the theory; indeed, this advance may be retarded by such ignorance. I am, however, more concerned with the wider effects of our neglect of foundations, the effects on the education of scientists, on general education, and on the whole standing of science.

With regard to the education of scientists, it is plain that the clearer the teacher, the more transparent his logic, the fewer and more decisive the number of experiments to be examined in detail, the faster will the pupil learn and the surer and sounder will be his grasp of the subject. It is true that if all teaching were exclusively of this kind the student would get a very wrong idea of the nature of scientific progress, but for the time being we are in little danger of falling into this particular trap. Our peculiar fault is that it seems to take one and a half to two centuries to achieve this clarification with the odd result that our students fondly (and, I think, quite wrongly) believe that science progresses from easy to difficult theories, that the concepts of Newton and Laplace were elementary, those of Einstein and Heisenberg complex. This leads occasionally to the fantastic notion of a general advance in human intelligence, since, to put it at its most ridiculous, every schoolboy seems to be able to understand what it took the greatest genius of the seventeenth century to discover.

In my own view, if 1 percent of the effort spent on physics were devoted to clarification, we could soon teach the basic concepts of quantum mechanics to the general run of nine-year-olds!

However much or little we may in fact realistically hope to achieve, it surely requires little argument to plead that a better understanding of logical and experimental foundations will help the training of scientists. This is equally plain for the teaching of science to non-scientists, and indeed a substantial effort has been devoted to this purpose in recent years. There is no doubt that considerable additional advances are possible and that it will take a great deal of work to further clarify the structure of physics and to identify the crucial areas where the greatest understanding can be conveyed with the minimum of technicality and thus the minimum of teaching time. To have all these advances permeate the school system will require an equally great and exacting effort.

Teaching is likely to be the field of the greatest rewards. It may in due course result in whole nations having some real understanding of science, in citizens who can make political judgments on matters involving science without the feeling that they are dealing with witchcraft. An essential part of such an advance lies in giving to the most intelligent and influential nonscientific members of the community a real respect for, and understanding of, the logical nature of science. At present, the attempts to teach modern science to such highly discriminating minds result only too often in their feeling that science is a mishmash of rationality and irrationality where logic has a very limited role. This causes many of the ablest youngsters not only to shun science as a career but to get the impression that the study of science is intellectually not quite respectable and involves much of a mindless grind that maims the thinking of all but the very best scientists who then, in pulling success out of this mess, appear more as magicians than as thinkers. The more we can strengthen the logical appeal of science, the more we can, at least in well-established fields, limit the technicalities necessary to understanding, the more successful we will be in making science look a great and intelligent human enterprise to the intellectually most capable.*

In recent years I myself have done some work in this direction on special relativity,[1,2] and in the far more difficult field of general relativity great advances in understanding are due to Fock's discussion of its foundations.[3] In this article I propose to sketch some thoughts on a few aspects of the foundations of quantum theory.

Traditionally, the aspect of quantum theory that causes the greatest conceptual difficulty to beginners, to the general public, and particularly to philosophers, is *indeterminacy*. Yet it seems to me that this idea was already latent in early thinking on the atomic structure of matter and became inescapable with the discovery of radioactivity which of course antedated Heisenberg's work by thirty years.

A mainspring of scientific endeavor has been the thought that, by digging beneath the appearances of our complex surroundings, we may find something simple. Such was the case with early ideas on atoms, the supposedly simple elementary bricks of matter whose complexity was then imagined to be "only" a matter of their arrangement (crystallography, molecular biology, and the theory of the liquid state are good reasons for putting quotation marks

* Much of what now comprises "popular science" in books and on TV does just the opposite by trying to evoke the response, "How incredible!" through showing dazzling successes of science combined with "explanations" that are almost designed to be regarded as incomprehensible and hence to be admired.

round *only*). The main supposed simplicity of this was the likeness of the atoms to each other. When Dalton made these ideas precise, he took it that all atoms of one and the same element were identical, a thought that, but for the notion of isotopes, has remained unchanged. This *identity* of the atomic species is, however, a deep idea in which lies the germ of many later developments. In everyday life we regard it as impossible that any two objects should be identical. However alike two golf balls or two coins or a pair of "identical" twins may be, we are wholly convinced that a sufficiently thorough investigation will necessarily reveal differences. Whatever the number of identical attributes, we feel sure to be able to find others which differ. In measurements yielding numbers that can take a continuous set of values (e.g. the weight of the golf balls), we feel certain that discrepancies will show up if the measurement is made precise enough. In the case of measurements yielding discrete values, though (the number of dimples on each golf ball), we give up when we have made sure that the two balls are the same as far as this number is concerned, and try other measurements to discriminate between them.

What, then, is the difference between Dalton's truly identical atoms and objects in everyday life? To start with, we must take it that a finite number of measurements establishes the identity of atoms, that we cannot go on endlessly making new and different measurements. This means that we assume that atoms possess only an *exhaustible* set of attributes, that only a limited number of measurements is at all possible on an atom (in my teaching experience students seem to be quite ready to accept that objects as difficult to observe as atoms should have this property but I cannot myself pretend that it is in any sense obvious). Moreover, where a measurement gives a number that in principle can take a continuous set of values (e.g. its mass or the electric charge of its nucleus), we must assume that for no evident reason the number "accidentally" always takes the same value for atoms of the same species so that no refinement of measurement can distinguish between them. On the other hand, we might suppose that, contrary to first appearances, this number can in fact, through the operation of some law of nature, take only a discrete set of values so that this property is indeed *quantized.* Thus the very notion of identity has far-reaching consequences which might be investigated further with advantage.

The notion of the identity of atoms is (and has for long been) so appealing because it seems to simplify the picture of the world and thereby to increase our knowledge. In fact, we can look at the idea in a different light. If we accept the idea that atoms are identical, we give up any idea of ever being able to identify them, of ever being able to pick out one atom out of a crowd of atoms of the same species. Therefore we accept with the notion of their identity a limitation of our potential knowledge and accept the notion of *indistinguishable objects.* This fundamental idea of quantum statistics is thus already inherent in Dalton's work. In a course on quantum theory, Bose-Einstein statistics could thus be derived at a very early stage.

The notion of identical indistinguishable atoms leads to most important consequences when radioactive decay is considered. The simplest experiments on radioactive decay show that, with regard to the timing of decay, otherwise identical atoms behave *differently.* Even a cursory examination of decay chains shows that atoms do not decay at the same age (and the idea of age as an attribute is anyway incompatible with the notion of identity), nor can the instant of decay be linked to any other of the exhaustible set of attributes of the atoms. Thus the notion of *indeterminacy* and of uncaused events is an inescapable consequence of the discovery of radioactivity for the theory of identical atoms. To put this a little differently, when a test is performed on a set of identical objects, either they respond identically, or they respond in different ways which can then only be *statistically distributed* in an individually unpredictable way. We thus arrive at the point so strikingly and clearly put by Popper[4] when he said that quantum theory gave statistical answers because it asked statistical questions. The prediction of nonidentical behavior of identical individuals can be carried out only in a statistical manner.

What interests me about this derivation is:

i. that it depends only on concepts and experimental results known to a very wide class of students;

ii. that it depends only on items known before 1900;

iii. that it gives a hint toward the quantization of charge and mass;

iv. that it says nothing whatever about the magnitude of the indeterminacy occurring.

It does not seem to me to be possible to derive the magnitude or indeed the dimensions of the uncertainty, that is, Planck's constant, solely from these ideas of atomicity and data on radioactive decay. The most direct way of arriving at this seems to be Einstein's work of 1905 on the photoelectric effect. It is interesting that while in most cases the original derivation differs from the easiest one (e.g. Michelson-Morley and radar in special relativity), in this case a very early paper is concerned with notions that can be simply presented. Its only predecessor was Planck's deep and fundamental paper of 1900 which cannot, however, be presented without a thorough discussion of blackbody radiation, or other subjects found difficult even by good students.

In using the data (and modern ones are clearer than those known to Einstein in 1905) on the threshold frequency and on the energy of the electrons released in photoelectric interactions, one gets, in the familiar way, the ideas of Planck's constant and of photons. With the notions of atomic indeterminacy already introduced, it is less difficult to appreciate that the grain of light and of atoms and nuclei is the same and corresponds to a *universal uncertainty*. Finally, I cannot see why the so-called wave-particle duality need cause as much consternation to students as is normally thought. It is after all clear to every student that in the most ordinary circumstances statistical quantities are different *kinds* of numbers from actually occurring ones. Everybody appreciates that the average number of children per family in a country is a *continuous* variable,* while the actual number of children in a family is necessarily an integral number. It is therefore in no way surprising that a different *type of number* is necessary to specify the probability of finding a photon from the type of number (discrete integers) required to count photons and hence interactions due to them. Of course, this argument makes it in no way clear why the statistical type of number should be based on a complex rather than a real and positive number; and I believe that there is no way of deriving this most fundamental result without referring to interference experiments, as Landé himself has done so beautifully. This remarkable fact must not, however, be mixed up with the obvious one that interactions are discrete while their probability of occurrence is not, and that thus the two must be described differently.

This brief contribution does not pretend to be a derivation of quantum theory or even a sketch of such a derivation. What it does hope to be is an encouragement toward the necessarily long and arduous establishment of a derivation based on a logical minimum of easily defined and crucial experiments together with an old familiar background of theoretical preconceptions. In addition to greatly easing the task of teaching quantum theory, it may even help in its understanding by its practitioners.

Such a program is probably unlimited. How simple are the experiments which may turn out to be all that is required? Could the quantal structure of matter be derived from the observation that there is a solid state of matter? Or, if this is too ambitious, from the fact that there are spectral lines?

The further we go in this direction, the more the experiments we are taking as basic are also the ones about which we learn in our

* To be precise, it is not continuous but discrete, the steps being the reciprocal of the large number of families in the country. This kind of situation is common. For example, the identity of species of atoms is only significant because the number of species is dozens of orders of magnitude less than the numbers of atoms, even of uncommon species, in the solar system, let alone the observable universe. Even here, though, the ratio is actually finite.

first few years of life. We then, for example, make our first contact (usually painful!) with matter in the solid state. It is these early experiences that shape our whole mode of thinking about physics, it is these experiences that form our common sense. The more the reputedly "mysterious" parts of physics can be shown to be logically linked to common sense, the better it will be. But even if progress in this work is not quite so spectacular, it will still be most useful.

References

1.
H. Bondi, *Relativity and Common Sense* (Heineman, 1964).
2.
H. Bondi, *Assumption and Myth in Physical Theory* (Cambridge University Press, 1967), chaps. 2 and 3.
3.
V. A. Fock, *Relativity and Gravitation* (Pergamon Press, 1963), chaps. 1 and 2. See also the first of his Copenhagen lectures.
4.
K. Popper, "Quantum Mechanics Without the Observer," in *Quantum Theory and Reality*, ed. by M. Bunge (Springer-Verlag, 1967), p. 15.

Seventeen **André Mercier**

Forms of Determinism, Objectivity, and the Classification of Sciences

Introduction

We know that in the limit of infinite light velocity, the kinematics of special relativity apparently goes over into Galilean kinematics, and classical mechanics applies. The relativity involved here is a relativity of space and time. In classical mechanics, time alone is used as the independent variable upon which all other physical magnitudes are supposed to depend, and for this reason time is never to be measured or observed; it is to be given. Hence, propositions in classical mechanics take the form: "At given time t, ..." Now, classical mechanics is known to be ruled by Laplacian determinism. This determinism can be fully appreciated when it is defined as the property of a theory to be able to predict, from given initial values, all the quantities characteristic of a system. In the Hamiltonian form of Newtonian mechanics, it is sufficient, as well as necessary, to give initial values of the positions and momenta of the particles of the system, because all other quantities are functions of the positions and momenta, and the fundamental equations allow us to solve for the positions and momenta as functions of time. No difference is made between c-numbers, q-numbers, and φ-numbers; but, in a sense, t is (physically) different from the other magnitudes. However, it should be noticed that Hamiltonian mechanics allows for freedom in the choice of such initial values at *any* time chosen as the initial time. This situation has led people to believe that Laplacian determinism truly describes the only conceivable cause-effect relation.

Actually, this is but one of the possible forms of determinism, called Laplacian. If Newton's original form of mechanics had postulated that forces determine the third, instead of the second, time-derivative of the position vectors of mass points, then we would have to count positions, velocities (momenta), and accelerations among the required initial values. This would not be traditionally Laplacian determinism, but it would be another form of determinism within a general idea of causality. In physics proper, causality always arises from the category of *causa efficiens.* Third-derivative determinism of the sort just described has never been envisaged as valid. Second-derivative determinism implies self-forces, which are in essence contradictions, when applied to electromagnetism.

But the question arises as to whether physics or science generally has considered, or should consider, other determinisms. The answer is in the affirmative. We can trace at least three main ideas of determinism and possibly suggest some more in connection with the classification of the sciences. One of them might be described as a superdeterminism, when compared with any other determinism, especially with Laplacian determinism; it is connected with Einsteinian relativity. Another one is sometimes, in comparison with Laplacian determinism or superdeterminism, called indeterminism by those who wish to be hypercritical of quantum theory. A further determinism is related to the general tendency toward greater entropy of isolated systems. *Causa finalis* has since the time of Aristotle been distinguished from *causa efficiens.* We do not contend, however, that they are irreducible; they are both causes and very "physical" ones indeed in the Greek sense, according to which physical implies all that is in nature. For this reason, we feel inclined to make no differentiation between efficiency and ends within a principle of causality, since ends are as efficient as anything else. Hence, there is a further form of "determinism" that is not mechanistic but vitalistic (final) in nature; if mechanistic and vitalistic views clashed in the controversies of former centuries, this was due to lack of insight. In the light of modern epistemology, they should be looked upon as compatible, although different.[1]

1. Superdeterminism

What this new word is meant to convey is implicitly contained in the Minkowskian space-time description but is usually not noted there, because most interpretations of the Minkowskian picture have run in the opposite direction. In epistemological foundations of special relativity, people will usually say that time is "added" to space in order to make a four-dimensional (pseudo-Euclidean) manifold. (In general relativity, this manifold becomes Riemannian.) The careless student will then tend to believe that "time has become a fourth dimension of space," meaning "physical" space, as if there were a "thing" (Latin: *res*) called real, physical space. But he is also reminded of the remaining difference between time and ordinary space by the signature $+++-$, and furthermore by the irreversible flow of time. The latter, however, is not quite correct, for in special relativity just as in Newtonian mechanics everything evolves reversibly, i.e. a "t and q" or a "t and p" theorem applies (classical precursors of the T.C.P. theorem).

What ought to be said, however, is not that t is made into a coordinate of a four-dimensional physical space, but that (ordinary) space with its three Cartesian coordinates has been assimilated to time. By this, we mean epistemologically that positional space has been robbed of its observability and converted into an independent system of variables together with (ordinary) time. The way this is done requires an explicit relation; this relation is *relativity* (special, then general), and it explains why Galilean relativity is not simply special relativity in the limit $c \to \infty$, but the case in which relativity in that new sense collapses, or does not arise at all. Initial values are no longer given "at an initial time." There are no initial times. There are world lines, each point of which is called an event. No mechanism working locally should be allowed for, though one does introduce forces artificially, as well as clocks and rods, leading to dilatation and contraction theorems. General relativity shows the restriction still more immediately. The only sensible question in general relativity theory would be to ask for the actual metric field of the universe. Once the answer is known, everything would be completely, that is, *super*-determined. Gravitation, if this interaction only is assumed, could be explained "without gravitation"

(J. A. Wheeler). Any unified field theory is a piece of wishful thinking meant to produce this same superdetermined picture of a universe, in which pre-Einsteinian observers had believed it possible to observe several interactions. In such a theory there is, in fact, no place for observation or measurements, if not by artifices designed to break the superdeterminism.

2. Laplacian Determinism

This concept, one would think, needs little comment, as everybody knows what it signifies or believes he does. But is everyone aware of the fact that classical mechanics makes a number of assumptions that remained tacit until its complete axiomatics became available?

A very important feature associated with Laplacian determinism is one which seems to be in contradiction to it, namely reversibility, meaning that a "t and q" or "t and p" theorem applies (as far as the Hamiltonian can be made invariant).

A further remark is important. Classical mechanics is fundamentally only a gravitodynamics. In other words, only gravitation as an interaction has its proper place in it. Of course, all sorts of other forces can be included, but either these are all artificial constructs or they rest on the introduction of material constants which, as in the case of elastic bindings, call for a further explanation. In particular, none of the other fundamental interactions (electromagnetic, weak, strong, etc.) can be properly dealt with by mechanics. Therefore Laplacian determinism need not apply to them.

What should be especially considered here is that Laplacian determinism allows for the possibility of a choice of initial values. We can, so to say, shoot at targets if we want to, whereas the superdeterminism of relativity theory does not give that freedom, unless we artificially break the cosmologic structure of the whole affected universe or consider simply those world lines that eventually go through the point events of interest. In quantum theory, there are also targets, but the experiments using them require a quantum-theoretical treatment. Moreover, the requirement that determinism applies, in the sense that systems should evolve in the direction of the future, would also break the structure of classical dynamics.

A special feature of gravitodynamics is that all its predictions are verified by observations that employ light rays, an intermediary which, in the Newtonian picture, is assumed to be totally insensible to gravitation and incapable of interactions with massive bodies. Light is the propagator of another form of interaction, namely electromagnetism. The only necessary assumption is that massive bodies reflect (or rather diffuse) light. This is the first reason why observers appear totally independent from the objects of their consideration; the second reason is that light does not make any noticable impact on the observers other than supplying the necessary information transmitted to the brain for mental comparison with the constructed predictions. This circumstance was influential in shaping the nineteenth-century concept of objectivity and then served as a paradigm for its further elaboration. As we shall explain further on, this was unfortunate, for objectivity need not take the particular form of independence of the subject from its object. After all, such objectivity is (approximately) realizable only if observations or measurements are made by means of signals that transmit interactions which are weak enough in comparison with, and irreducible to, the specific interaction involved in the definition of the system under observation.

3. Thermodynamic Determinism

Irreversibility in thermodynamics or the H theorem in statistical mechanics have never been associated with determinism in the literature. This is understandable insofar as irreversibility does not involve a precise cause for a definite effect, and hence its description is not based on such exact data. If such were the case, thermodynamics would rest on differential equations instead of, as in reality, on differential "inequations." However, the latter are a kind of generalization of equations, and in this sense we feel justified in saying that thermodynamics does treat a form of determinism in that it accounts for the direction taken by the evolution of systems from a state of disequilibrium toward a state of equilibrium.

This third kind of determinism has the effect of showing, like the aforementioned ones, that every kind of determinism depends on the choices offered by each theory in the description of the initial states of the physical systems under consideration. Super-determinism leaves no choice, unless it is broken and the cosmological view replaced by a partial or local view, attached to a physical (not merely mathematical) frame of reference. Laplacian determinism does allow for a choice, which is best carried out among canonical variables, but we can lower the precision by considering merely a distribution density in phase space obeying Liouville's theorem.

However, these two cases are—with or without breaking or weakening the determinism—described by regular differential equations. Thermodynamic determinism is quite far removed from such a description and is therefore perhaps already outside the scope of ordinary physics, although many physicists would not hesitate to integrate thermodynamics into physics. Yet apart from the fact that, as Stueckelberg insisted, (reversible) physics should be considered an approximation of thermodynamics, whereas thermodynamics is in no sense a special case of physical theories, thermodynamics is fundamentally different from the rest of physics; for it deals with engines and not with mechanisms. Or, to put it more philosophically, it deals with a temporal category of a kind that is irreducible to the temporal category associated with the study of interaction proper. (Mechanisms are models of interacting systems; engines are not.)

4. Determinism of Ends

Many a philosopher will find this designation a misnomer, for determinism and finalism have been considered antinomic. First, however, the antinomy arose in former days, not between determinism and finalism, but between what was called mechanism and vitalism. Second, the understanding of the one-to-one connection between the notions of mechanism and interaction has rid the concept of mechanism of its connotation as a doctrine. The same must happen to the notion of finalism: finalism should not denote a doctrine, but *a* finalism should be conceived as the model of an organism which tends in a specific manner toward ends.

Engines do not work toward ends; their workings are ruled by thermal evolution, according to which they dissipate energy. Mechanisms neither move toward ends nor do they dissipate energy; each mechanism can be defined as a clock (i.e. a periodic phenomenon). A pure mechanism is not subject to thermodynamic determinism, and a pure engine is not subject to "finalism" (to determinism of ends). But a finalism involves one or more engines, and an engine includes one or more mechanisms.

5. The Arrow of Time

In (at least) two respects, mechanisms seem to be nevertheless related to irreversibility. One concerns the cosmological evolution of the world, at present believed to be an expansion in the course of cosmic time. Cosmic time is a possible construct in Einsteinian cosmology, and expansion is a possible choice among the solutions of Einstein's equations. The other respect in which a mechanism seems related to irreversibility concerns the connection between (electromagnetic) waves and their sources. Its description is usually believed to be a retardation, which is one possible choice among the solutions of the wave equation. A relation between this choice and the thermodynamic arrow of time can be postulated; no existing theory can lead to such a conclusion from general principles.

The following considerations, however, do not seem totally absurd. (a) In a way, anything which can be said to be a propagation is somehow a wave propagation. (Whether gravitational waves exist may be debatable, but the whole gravitational character of the Einsteinian universe can be split into news functions traveling wavelike from body to body.) The same applies to the electromagnetic field and essentially to any field (whether quantized or not, which is irrelevant here). Maybe one should not ask whether retardation (as a choice) creates an arrow of time coinciding with that of irreversibility, but rather say either that the irreversible character of thermodynamic determinism requires (the choice of) retardation, which would apply to all waves, or at least that it requires one definite choice, excluding the other, for we should not forget that no system in nature has yet been found to be as strictly elementary as a point-mass should be. (Even "elementary" particles may be made of quarks, and quarks will presumably split into subquarks in the year $X > 1970$.) So everything is composed of other elements; hence everything is a statistical ensemble subject to the H theorem. (b) The universe is the totality of everything. Increase of entropy applies strictly only to the universe as the one completely isolated system. Again, rather than saying that expansion creates an arrow of (cosmic) time that coincides with that of thermodynamic evolution, one should perhaps say that the inevitable increase of entropy of the universe requires a choice, for example, expansion, in the evolution of the universe.

Such requirements must be made in order to subsume the principles of the (less comprehensive) physics of reversible processes under the principles of the (more comprehensive) theory of irreversibility.

6. Quantum Determinism

Having explained the notion of determinisms, we can now move on to quantum theory. Quantum determinism is intended as a name for a feature possessed by current quantum theory. But what is "current" quantum theory? We do not pose this question with reference to "interpretations" (so-called Copenhagen, so-called hidden variables, etc.). We raise it in the following context: Is quantum theory a general frame for theories of various interactions, or is it the specific theory or dynamics accounting for one particular kind of interaction, namely electromagnetism? In other words, is quantum theory the form best suited for a physical theory at the present time, or is it strictly only (quantum) electrodynamics? There was a time (up to 1905) when Newtonian (or the more elaborate Hamiltonian) mechanics was regarded the best form for any theory. Today, it must be viewed as a good approximation of gravitodynamics, whereas Einsteinian gravitodynamics is a still better approximation. But Newtonian gravitodynamics is in no case an electrodynamics, and originally Newton intended it to be nothing but a gravitodynamics.

Quantum theory was, to begin with, solely an electrodynamics; for atoms, molecules, and crystals are composed of electrically charged particles, while blackbodies involve additionally electromagnetic radiation. But, owing to a success reminiscent of the success obtained by Newtonian mechanics from the eighteenth century on, and also owing to the fact that strange (meaning alien to electromagnetic structures as such) features, like spin, were apparently finding their neat account within the framework of quantum theory, this theory soon gained acceptance as the general theoretical framework for physics. Fields of all kinds were "invented," "quantized," and even experimentally "observed" as mesons and other particles. Finally all imaginable fields, including even gravitational ones, were assumed to be quantizable. (We ignore the difficulties about nonlinearity.)

Now, on the one hand, electrodynamics must be elaborated on a relativistic basis. Therefore quantum theory, if it is to be, or to include, quantum electrodynamics, must be made relativistically invariant. One might therefore expect it to obey superdeterminism. On the other hand, we know of Heisenberg's uncertainty relations, of de Broglie's wave-corpuscle duality, of Dirac's nonsimultaneity between canonically conjugate observables—in short, of Bohr's complementarity notion, sometimes commented on with the help of a concept of indeterminism. In our opinion, indeterminism is a bad appellation for what applies in quantum theory, as it is really a specific form of determinism. It is also wrong to say that quantum theory is an acausal theory, for quantum theory is concerned with the predictions of effects due to causes. (It speaks about the photoelectric effect, the Compton effect, etc.) It makes essentially two kinds of predictions: it allows us to calculate stationary states, and it predicts the probability of a system either to be found in a particular state, or to jump from an initial state (in-state) into a final state (out-state). Its predictions are extremely well verified by experiments (essentially measurements of spectra and cross sections). These predictions are always made on the basis of given initial conditions. Of course, these initial conditions differ in principle from those used in other determinisms, since they must be chosen out of a complete set of proper states all belonging to a set of commuting operators. Therefore, quantum theory deals with a "conditional" determinism as compared with Laplacian or superdeterminism.

How is it then to be understood that quantum theory employs either coordinates which are Cartesian and imply the existence of (at least symbolically) objectively independent observers, or momenta obtained from the coordinates by Fourier transformation, and moreover has to include electromagnetic fields and charges? The use of coordinates or momenta, i.e. of observers endowed with some sort of objective independence, may not be in contradiction with quantum determinism; however, the reference to electromagnetism seems to require superdeterminism on account of the relativistic covariance of fields and charges.

How do we resolve the apparent paradox? To answer this question, it must first be said that the axiomatics developed so far within the quantum-theoretical framework for the clarification of quantum observability tacitly refers to nonrelativistic quantum theory. Furthermore, we must recall, in agreement with V. Fock, that covariance and observation are two distinct matters. Quantities belonging to a system can be covariant with respect to particular coordinate transformations; this is a mathematical property. But by choosing the particular (Galilean, Lorentzian, or more general) covariance underlying a theory, one has not yet said anything about the way properties of the systems under consideration are measured. Measurements require apparatus which are made of material elements. They must interact with the observed system; otherwise no change occurs that would be noticeable by counting events, reading scales, or the like. Superdeterminism applies if measurements are made impossible by the fact that interaction can be switched neither on nor off.

To be precise, no measurement is allowed if relativity theory is conceived of as a theory of the physical world rather than merely a mathematical framework; for, in this view, each piece of matter has (or rather *is*) a world line in space-time, purely and simply. To express it colloquially, nobody could, according to this superdeterministic theory, deposit money in his bank, for he disposes of nothing that is called money; he is himself a world line, and there is no such thing as a bank which he could choose to enter or not. There may be bank notes, but their world lines are unchangeable. Covariance warrants only the invariance of these world lines and of other properties in case the parameters used in their description are exchanged against others. It prescribes further how properties change in mathematical appearance if such exchanges take place.

Since, in fact, superdeterminism excludes measurements, it renders physics impossible, unless either the universe is empty, or physics is identical with a closed cosmology of the universe as a whole. In either case, there can be no objects to observe and no subjects to do the observing. The question whether the theory is objective does then not arise.

All other determinisms each imply one notion of objectivity based upon the distinction between subjects and objects or, in more technical language, between measuring apparatus and systems. But some identification is then to be made between the measuring apparatus (the subject) and the parameters used to describe the system (the object). The identification must include the choice of the covariance (whether Galilean, Lorentzian, etc.), but it will still leave open the conditions under which measurement is allowed.

In thermodynamics, for instance, the covariance is based upon the assumed existence of an equation of state, and the system must be in thermal equilibrium in itself and with the measuring apparatus for a measurement to be possible. The phrase "increase in temperature," for example, has no real meaning, for one has to wait an infinitely long time until a temperature is readable (practically, ten minutes will suffice for patients in the hospital). Moreover, the reading is made by use of light signals alien to the definition of a thermal state. This amounts to a conversion of the system into an isolated system by inclusion of the apparatus, which is then a reduced or partial cosmos, subject to its own kind of superdeterminism. Thermodynamics is thus a theory of how isolated, superdetermined, partial universes can be broken up and changed into others; this is how thermodynamic determinism enters the picture.

In the Newtonian theory and specifically in Hamiltonian dynamics, measurement is, in the first place, possible under any circumstance describable within the frame of the theory; this means that no choice is necessary of those properties that are to be measured to the exclusion of certain other ones. Second, the intervention of measurement leaves no trace upon the system itself; only the measuring apparatus is affected—momentarily, by illumination of a scale, or permanently, by a trace on a photographic plate or information in the brain.

According to quantum theory, measurement is also possible under any circumstance describable by the theory, but a choice of observables has nevertheless to be made; yet not only is the measuring apparatus changed (e.g. by a trace on a photographic emulsion), but the object observed itself experiences a mutation of its state. This is catastrophic in a double sense: (a) in comparison with the object's "size" (e.g. its momentum or energy content) and (b) by rendering this object unobservable a second time along the same evolutionary track. For this reason, the switching on, and then off, of the interaction between subject and object can occur only once.

This sort of description shows that one should not be misled by vocabulary. In the early days, Heisenberg's uncertainty relations were sometimes summarized as a "principle of indeterminacy," and commentators drew a conclusion of indeterminism, eventually describing quantum theory as devoid of causality or as contradicting a "principle of causality." This amounts to a hoax. Bohr himself has explained at length that the very idea of complementarity is meant, among other things, to save causality. But the workings of causality according to quantum theory are different from its functioning in other determinisms. Those investigators are on the wrong path who search for logical or material errors in current quantum theory in the belief that quantum theory is fundamentally not deterministic (meaning usually Laplacian, not superdeterministic, a term which has not yet been properly analyzed in the literature). We have no objection to attempts to construct another theory that could imply a new determinism, unthought of as yet. But quantum theory is perfectly sound as long as we neither seek things in it that are not there nor put things into it for which there are no places.

Finally, covariances, often referred to as invariances, and determinisms need to be distinguished, as has been suggested. They have separate axiomatic systems. By the way, the axiomatics of modern quantum measurement demonstrates that the ideas of "initial state" and "state at time of observation" have to yield to the concepts of "in-state" and "out-state" in the sense in which these are used in dispersion theory or S-matrix theory, where the initial state is a state at time $-\infty$ and the state for the completed observation is at time $+\infty$. This temporal aspect was more or less included in old-fashioned quantum theory, which required that long enough time intervals should elapse between the switching-on and switching-off times of the perturbation; however, these were also assumed to be not too long if Born's approximation formula was to apply.

As an aside, let us recall that quantum theory must postulate a measuring device (subject) distinct from the system under observation (object) and differing from the latter by being of so-called macroscopic nature, so that it has the same intrinsic properties as measuring devices within Laplacian determinism. This distinction implies in particular that there can be no quantum-theoretical treatment of the whole universe: the whole universe is never a quantum-theoretical system; there is no quantum cosmology.

However, the question whether quantum theory is merely an electrodynamics or whether it is the (most advanced known) theoretical framework for physics is still unsettled today. Since Lorentz covariance is possibly valid more generally than just for electrodynamics—for c is not merely the velocity of light but the limiting velocity which distinguishes tachyons from other (ordinary) particles—one is inclined to believe that quantum theory is, at least to a good approximation, more than mere quantum electrodynamics. In any event, quantum electrodynamics in Lorentz-covariant form is not free from divergences that can be avoided only artificially and that seem to arise from the simultaneous application of both quantum-theoretical and Lorentz-covariant conditions. These difficulties appear in any relativistic quantum field theory. Perhaps quantization and Lorentz covariance (and still more general covariances) are in a way incompatible, even if it is possible formally to write quantum theory in a (Lorentz or general) covariant fashion; and it is indeed extremely difficult to quantize generally relativistic gravitodynamics. Maybe this alleged incompatibility is caused by the superposition of two irreducible determinisms. But these are speculations.

7. Relativity of Subject and Object

At the beginning of this article, we recalled the relativity of space and time and then showed how it affects the determinism involved. At the end of Sec. 6, we found that the adoption of a particular determinism is dependent upon the view taken toward measurement, meaning the assumed relation between what we symbolically call subject and object. But at the same time, the relativity postulated between space and time furnishes the fundamental structure upon which the definition of covariance is based.

Consequently, there are two kinds of relativity: the relativity conceived along the lines of the Galilean-Einsteinian tradition and the relativity conceived between subject and object. In one case, the relativity adopted defines the kind of space used to describe temporal behavior of material systems, in the other case, the relativity assumed determines the kind of device used to verify "objectively" the description made. There is then indeed relativity in so-called objectivity; and it is questionable whether one should stick to the classical concept of objectivity, which was unconsciously adopted by scientists and philosophers from a tacit acceptance of Laplacian determinism, to which they attached universal validity. Instead, objectivity should be looked upon as a general idea proper to science and to the pursuit of truth, allowing the relativity between subject and object to be adapted to the various possible determinisms.

8. Determinisms and the Classification of Sciences

From a methodological point of view, different determinisms may be regarded as defining different approaches to scientific problems. At the same time, they seem to disclose different ontological situations, although we do not claim to have proved this. We do not know either whether the remark made about the difficulties around diverging self-energies is well founded; but if it is, then this example suggests the possibility that the combination of two approaches, each having its own determinism, may produce an epistemological situation considered most unsatisfactory by specialists: everyone looks for artifices that would allow the elimination of divergences. The superposition of two such determinisms might be paradoxical. Perhaps paradoxical situations should be avoided, but they are not necessarily contradictory.

The various determinisms, defining various scientific approaches, can be thought of as occurring at different levels. The upper levels would assume the existence of lower ones, but an activity developed at one level need not take lower levels explicitly into account; though it may be instructive, useful, and even desirable to superpose considerations from adjacent levels, e.g. by elaborating a relativistic quantum theory or by assuming van der Waals forces between molecules in order to establish an explicit equation of state. The superposition of all the determinisms indicated so far, including finalism, would lead to a comprehensive biophysics.

These levels allow us to move from one subdivision to another, or even from science to science. Roughly speaking, we successively encounter physics (proper), thermodynamics, biology, and other sciences. It would be a worthwhile task to analyze biology and even other disciplines like psychology and sociology in detail, just as we have done for physics, that is, according to the various "determinisms" involved at these further levels.

This clearly suggests the possibility of a classification. Other classifications, e.g. the positivistic one, have been proposed on the basis of the subject matters of the various disciplines; for instance, physics deals with matter, biology with life, psychology with the soul, etc. But what are matter, life, soul, etc.? Modes of determinism as criteria for a classification offer a double advantage: they admit of rather clear definitions, and they can be well distinguished from one another.

The question as to whether they do correspond to different ontological situations or are of purely methodological interest will not be discussed in this context.

Reference

1.
See in particular A. Mercier, *Fugit irreparabile tempus,* Berner Rektoratsreden (Verlag Paul Haupt, 1967).

Eighteen **Paul Bernays**

Causality, Determinism, and Probability

I am aware of the difficulties inherent in fulfilling Wolfgang Yourgrau's friendly request to contribute to the collection of articles in honor of Alfred Landé, inasmuch as I cannot lay claim, in contrast to most of the other contributors, to any particular authority in Landé's main field of investigation: quantum mechanics. This circumstance would argue against my contributing if it were not for the fact that in the discussions of quantum mechanics physicists find themselves in polemical realms. Here not only are their views not determined by means of experimental and mathematical results and previous experience in the physical sciences, but they are to a large extent influenced by philosophical considerations, in particular those of positivistic philosophy. The leading physicists have fundamentally different opinions as to how quantum theory should properly be interpreted, and it is precisely to questions raised in this connection that Alfred Landé's more recent publications are addressed. Since I belong to those who welcome the impressive criticism presented in Landé's writings of sundry prevailing views, permit me then to give vent to my opinions concerning certain points of the topic under debate in terms of general considerations of methodology. These considerations concern the relationship of causality, determinism, and probability.

1. Causality and Determinism

A traditional view is that the idea of causality becomes precise in a specific context only in terms of physical determinism. Determinism is connected to the mathematical form of physical laws, in particular those of classical mechanics, according to which the time-dependent derivatives of the state parameters of a system are uniquely determined by the actual values of these quantities. Consequently (upon appropriate mathematical conditions), the values of the quantities of state are determined for arbitrary times if they are given for some initial time. A physical law satisfying these conditions is called a deterministic law. A way of extrapolating the statement of the deterministic character of the laws of mechanics is the *deterministic doctrine* according to which one imagines all physical events in the universe to be within one total system whose stages of development form the solutions of a system of differential equations of the type referred to. It appears difficult to bring this doctrine into agreement with a probabilistic theory of the physical processes; hence the adoption of probabilistic methods is incompatible with determinism. Furthermore, this incompatibility argues against causality in general.

In dealing with these disputes, we have first of all to distinguish clearly between causality and determinism. The tendency of combining the two is due to the idea that determinism is the correct sharpening of the causality considerations. A view of this kind is to be found in Kantian philosophy according to which the Newton law that force equals mass times acceleration constitutes the time-dependent schematization of the general causality law for physics. This Newton law is interpreted to say that the force at any one instant effects a change in the state parameter—velocity. This interpretation, however, is thrown into question by the recent advances of physics in Einstein's gravitational theory. For example, according to this theory a planet describes a geodetic line in a gravitational field determined by the sun. And one should interpret this geodetic line (which is a generalization of the linear uniform movement) as expressing a persistence rather than a change. It follows that the planetary motion exhibits what may be called a "stationary behavior." Yet the causality principle finds application directly where a deviation from stationary behavior occurs.

In any case, the causal view is not bound to the theoretical scheme of mechanics or generally to the assumption of deterministic lawfulness. Indeed, the causal aspect by itself does not include the element of strict generality and of mathematical preciseness which is peculiar to deterministic reasoning. One might even observe that deterministic consideration tends to eliminate the proper causal aspect. In fact, in the presentation of a physical theory the concept of causality has no proper place.

The causal view has its importance:

1. in the *heuristic* reasonings of natural science,
2. in the establishing and correcting of experiments,
3. in the *applications* of physical theories,
4. in our familiar instinctive dealing with concrete events,
5. in most of our common reasoning about living nature, and in particular on human actions, motives, characters, and faculties.

By considering all these cases, one will find that the assumption of a strict determinism occurs in none of them. What we often need instead is practical certainty, whereas the deterministic form of differential equations can, from the causal point of view, be regarded as a mathematical idealization.

That the doctrine of determinism might be necessary for the causal way of thinking is not at all the case. Besides, this doctrine is not in the least justified from the standpoint of theoretical science. The deterministic form exists only for the theoretical description of a fixed, definite system. Such a theoretical description, however, can only schematically represent any *Gegenstandlichkeit*; and the manner of schematization depends essentially upon the *order of magnitude* of whatever is under consideration. In the course of many astronomical investigations the earth can be represented by a point; naturally, in the investigation of the operation of machines this is not feasible. Here again some factors have to be neglected which in the investigation of molecular processes would have to be taken into consideration. The idea of a total physical system embracing all phenomena is due to a neglect of the essentially schematic and approximate character of the way of description in theoretical physics. There is no kind of approximation appropriate to every object of investigation.

2. Causality and Probability

Our considerations up to this point have served to demonstrate that causality is in no way necessarily connected with determinism.

With the differentiation between causality and determinism in mind, we further come to see that the application of statistical and probabilistic methods in physics is not in conflict with the causal view. Rather, in reference to the probabilistic and statistical approach to causality the following remarks seem to be in order:

a. The applicability of statistical and probabilistic methods to a multiplicity of processes will not be impaired if we assume that the individual processes are causal. Thus, for instance, no argument against the usefulness of accident statistics for the purposes of insurance evaluation can be made from the circumstance that the causes of individual accidents can occasionally be found.

b. In advancing statistical estimates and probabilistic considerations, one must continually take into account the causally relevant conditions. We can expect a mortality table to remain applicable only if the conditions that apply to a certain population (nutrition, the state of knowledge in medical science, the political situation) remain approximately the same. In a probability evaluation of the distribution of water drops on a surface, one must consider whether the liquid issues from a source in divergent, or approximately parallel, jets. Deviations from results anticipated by probability theory (in virtue of certain assumptions) could also lead to the discovery of causally modifying circumstances.

c. In many cases a statement that some circumstance is the cause of an event does not mean more than that an overwhelming probability for the event is produced by that circumstance. Thus, when a malady is said to be the cause of a death, then in general the lethal effect of the malady cannot in advance be predicted with full certainty; it is the course of the malady, depending on various details, that is finally deciding.

d. The concept of probability, as it is chiefly used in physics, contains either explicitly or implicitly the factor of causality. This factor appears explicitly when one characterizes, as Popper does, the probability as propensity. However, the causality factor is also essential in dealing with the frequency definition of probability, as it has particularly been worked out by von Mises. The term "frequency definition" gives rise quite easily to the impression that one is dealing here only with empirically determined quantities. In fact, though, the limiting value of the relative frequency* is not attributed to a single series of experiments—it arises rather from a "*Kollektiv*," that is, from a complex of conditions which are either at hand, or established, or assumed. Further, the correct view is that the results of a series of experiments made upon a *Kollektiv* express the *nature* of this *Kollektiv*. Otherwise, the series of experiments would have nothing to teach us about further series of experiments with the same *Kollektiv*.

The view we have presented of the close connection of probabilistic reasoning in physics with causal thinking seems to contradict Landé's view, since he expressly states "that there is no causal explanation for the harmony with statistical dispersion and mathematical random theory, hence that 'statistical cooperation' must be accepted as a basic feature in the world in which we live."[1] "Every statistical ensemble," he says, "shows that we may regard acausality (or unpredictability, using a subjective term) as an autonomous principle of physics."[2] However, it is to be noted that Landé speaks here about causality and causal explanation in the sense of the deterministic pursuit of individual events. The concept of causality used by Landé is indeed adopted by most physicists and also identified in the philosophy of Kant with the familiar causality concept. Our preceding reasoning suggests that it is suitable to distinguish the commonly accepted notion of causality from its mathematical "padding" of determinism, and to speak of causality only in the familiar, unadulterated way.† Doing this, we are also

* We do not regard here the difficulty connected with the limit.

† By keeping to the familiar concept of causality we certainly differ from the view of Max Black mentioned by Landé (Ref. 2, p. 26) that "Cause" acquires a meaning "only as a conscious and voluntary act of a subject making something happen to an object." Consciousness and intentionality set the standard where one is dealing with questions of guilt or of merits—not, however, for causality. When, for example, a man is struck dead by a bullet, it is indeed irrelevant with regard to *causality* whether the bullet was shot consciously and willfully or whether it was fired unintentionally.

able to give that basic property of our world, which is denoted by Landé as "statistical cooperation," a somewhat less paradoxical name, and call it instead the "causal relevance of probability." (It is this feature of our world to which Edgar Zilsel pointed in his book *Das Anwendungsproblem*,[3] exhibiting it as the common basis of the heuristic procedure of induction in empirical science and of the concrete applications of probability theory too.)

Likewise, what Landé states as a principle of acausality can be expressed by saying that every deterministic model in physics is only approximately true. This statement again can be drawn from the more general principle of Ferdinand Gonseth's philosophy that in our exact theories we cannot pretend to attain a fully adequate representation of reality but only a (more or less strong) schematic approximation. Thus, there is in fact no disagreement between the view we present and Landé's formulation; only the terminology we recommend differs from his. We shall also assent to Landé's thesis that probability is in principle the same in quantum theory as in prequantal physics.

Nevertheless, one striking difference emerges between the two in dealing with probability: Whereas in prequantal applications of probability theory to theoretical physics fundamental probabilities (that is, the basic suppositions concerning equal probability) are drawn either from relations of numbers or from relations of measure in Euclidean space, or those of a phase space, in quantum theory one has specific fundamental laws of probability. One is dealing here with the probabilities of transitions from states of elementary particles, which in some respect are differentiated in species $A_1, A_2, \ldots,$ into states which in some different respect are differentiated in species $B_1, B_2, \ldots$ (where the differentiations are in each case characterized experimentally). What is peculiar here is that the probability matrix P_{AB} with components $P(A_k, B_j)$—which represent the probability of transition from a state A_k to a state B_j and consequently are nonnegative real numbers—is reduced to a unitary matrix ψ_{AB} with (in general) complex components $\psi(A_k, B_j)$, by virtue of the relation:

$$P(A_k, B_j) = |\psi(A_k, B_j)|^2.$$

The founders of quantum mechanics have given this peculiar law a kind of physical interpretation which makes it necessary, however, to dispense with an unequivocal theoretical description of the quantum-mechanical states and processes. As a result, even the notion of physical reality becomes questionable. In the light of these consequences, Landé prefers to proceed according to the principle *hypotheses non fingo*. In place of a physical interpretation he gives a *mathematical motivation*, by showing that:

1. The aforementioned relation of the matrix P_{AB} to a unitary matrix ψ_{AB} constitutes the simplest (if not the only) mathematical possibility of fulfilling the basic conditions for the probabilities

$$\sum_j P(A_k, B_j) = 1,$$

$$P(A_k, A_i) = \delta_{ki} = 1 \text{ for } k = i$$

$$= 0 \text{ for } k \neq i,$$

together with the condition of symmetry*

$$P(A_k, B_j) = P(B_j, A_k),$$

and of assuring further that the combination law of probabilities will be obtained from a group-theoretical composition.

2. With the addition of a relatively simple assumption, it is possible to determine the form of the function ψ and from this function to arrive mathematically at the basic laws of quantum mechanics.

3. Causality and Time Direction

That probabilistic and causal views are not in opposition to one another, but rather that the probabilistic conception is a kind of

* It does *not* follow from this condition of symmetry that the matrix P_{AB} must be symmetrical; indeed, the condition that $P(A_k, B_j) = P(B_j, A_k)$ does not (in general) entail that $P(A_k, B_j) = P(A_j, B_k)$.

causal interpretation, is made notably apparent by the fact that for both the causal and the probabilistic explanation one, and only one, direction of time is distinguished, while deterministic laws, particularly those of mechanics, are characterized by reversibility in time. The familiar causal interpretation explains an event *always* in terms of something which occurred earlier, never something later.

In probabilistic reasoning, the direction of time is also essential. Many events in our familiar experience would become extremely improbable if reversed in time, for instance, the reversal in time of the process whereby a house is blown down by a wind and transformed into a scattered heap of stones. Another example of such an event, considered (in another connection) by Landé, is the following: balls of small radius issue from a pipe onto the edge of a knife, the back of which rests on a plate. Through the impact with the knife edge, they are scattered right and left upon the plate. Let us also assume that the plate is covered with some soft material so that the balls remain in place. By a reversal of this event the balls would have to move in a direction toward the knife edge in such a manner that in bouncing off the knife edge they enter the mouth of the pipe. This would be an instance of extreme improbability. Just as astounding and improbable would be the reversal of the event in which ink flows from a bottle, making a stain on a tablecloth.

Many physicists tend to regard the distinction of a time direction as something anthropomorphic—something that is characteristic of our experience but not of nature objectively. However, it should be noted first that the time direction of our experience is the common time direction of all organic development, and thus must be more than just subjective. Moreover, among the purely physical processes there are many sorts for which one and the same time direction is specific. To this group belong all the processes of diffusion, of explosion, of energy radiation, of heat conduction, as well as those just mentioned events whose reversal in time would make them highly improbable.

The common property of these processes is what physicists call irreversibility. When dealing in physical theory with irreversibility, a profound, systematic difficulty is encountered in the circumstance that the elementary laws of physics all have the character of reversibility; and this even holds for the basic laws of the transition probabilities. There was, of course, in the classical thermodynamics of Clausius the principle of entropy, according to which entropy (in a closed system) can never decrease in time but only increase (if not remaining constant). In the case of microphysics, however, where entropy is defined statistically, the principle of entropy is no longer regarded as being strictly valid. Landé concurs with this view, and he expressly maintains that "neither classical nor quantum theory permits a general entropic definition of the direction of time."[4]

Now, it is certainly not necessary that the direction of time be generally definable by means of a quantity of state. Nevertheless, the requirement to render understandable the essential role of the time direction for any probabilistic explanation remains. Landé adopts an argument adduced readily by physicists: "We happen to live in a very improbable state of the world which develops into a more probable state." We can sensibly speak of an improbable state of the world, however, only with respect to some *definite* order of magnitude of the description. Yet probabilistic assertions are in fact made with respect to *various* orders of magnitude. For each of them, it is the same direction of time which assumes the said distinctive and relevant role for a probabilistic understanding.

Now Landé takes a much better position here than those physicists who tend to regard the probabilistic method in physics as something purely mathematical. He is indeed acknowledging that the applicability of mathematical random theory to physics has to be attributed to a "basic feature in the world." From this point of view there is nothing shocking in admitting that probabilistic reasoning in physics has to be restricted by a condition concerning time direction. It might even seem that such a condition is actually observed instinctively and implicitly by the theorists.

The suggestion we have just made may be considered in connection with our first one: to separate causality from determinism and, furthermore, to interpret Landé's principle of statistical cooperation as expressing causal relevance of probability. This relevance, however, appears to have meaning only with respect to a distinguished direction of time—the same direction to which we commonly refer in causal explanation.

References

1.
A. Landé, *From Dualism to Unity in Quantum Physics* (Cambridge University Press, 1960), p. 8.
2.
A. Landé, *New Foundations of Quantum Mechanics* (Cambridge University Press, 1965), p. 32.
3.
E. Zilsel, *Das Anwendungsproblem* (J. A. Barth, 1916).
4.
Ref. 2, p. 75.

A List of Publications by Alfred Landé

Zur Methode der Eigenschwingungen in der Quantentheorie, *Inaugural Dissertation* (Göttingen, 1914).

Quanteneffekt im Hochfrequenzspektrum, Phys. Zs. **15**, 793 (1914).

Zur Theorie der Helligkeitsschwankungen, Phys. Zs. **15**, 946 (1914).

Einige neue Experimente zur Quantenhypothese und deren theoretische Bedeutung, Naturwiss. **3**, 3 (1915).

Die Beugung endlicher Wellenzüge an einer Halbebene, Ann. Physik **48**, 521 (1915).

Über ein Paradoxon der Optik, Phys. Zs. **16**, 201 (1915).

Theoretisches über die Breite der Spektrallinien, Phys. Zs. **16**, 313 (1915).

Die Abzählung der Freiheitsgrade in einer Elektronenwolke (strahlender Körper), Ann. Physik **50**, 89 (1916).

Die Abstände der Atome im Molekül und im Kristall (with M. Born), Naturwiss. **16**, 496 (1918).

Das elektrostatische Potential des Flusspatgitters, Verh. Deut. Phys. Ges. **22**, 217 (1918).

Über die Berechnung der Kristalleigenschaften mit Hilfe Bohr'scher Atommodelle (with M. Born), Preuss. Akad. **45**, 1048 (1918).

Kristallgitter und Bohr'sches Atommodell (with M. Born), Deut. Phys. Ges. **20**, 202 (1918).

Über die Berechnung der Kompressibilität regulärer Kristalle aus der Gittertheorie (with M. Born), Deut. Phys. Ges. **20**, 210 (1918).

Die Randbelegungsmethode zur Lösung von Potential- und Schwingungsproblemen, Ann. Physik **57**, 519 (1918).

Über die natürliche optische Aktivität isotroper Flüssigkeiten, Ann. Physik **56**, 225 (1918).

Über die Koppelung von Elektronenringen und das optische Drehungsvermögen asymmetrischer Moleküle, Phys. Zs. **19**, 500 (1918).

Dynamik der räumlichen Atomstruktur, Deut. Phys. Ges. **21**, 2, 644, 653 (1919); **22**, 83 (1920).

Adiabatenmethode zur Quantelung gestörter Elektronensysteme, Deut. Phys. Ges. **21**, 578 (1919).

Elektronenbahnen im Polyederverband, Preuss. Akad. **5**, 101 (1919).

Eine Quantenregel für die räumliche Orientierung von Elektronenringen, Deut. Phys. Ges. **21**, 585 (1919).

Das Serienspektrum des Heliums, Phys. Zs. **20**, 228 (1919).

Würfelatome, periodisches System und Molekülbildung, Z. Physik **2**, 380 (1920).

Über die Grösse der Atome, Z. Physik **2**, 191 (1920).

Über ein dynamisches Würfelatommodell (with E. Madelung), Z. Physik **2**, 230 (1920).

Störungstheorie des Heliums, Phys. Zs. **21**, 114 (1920). (*Habilitationsschrift*, Frankfurt.)

Über Würfelatome, Phys. Zs. **21**, 626 (1920).

Über die Kohäsionskraft im Diamanten, Z. Physik **4**, 410 (1921); **6**, 10 (1921).

Über den anomalen Zeemaneffekt, Z. Physik **5**, 231 (1921); **7**, 398 (1921).

Über den anomalen Zeemaneffekt, Naturwiss. **9**, 926 (1921).

Anomaler Zeemaneffekt und Seriensysteme bei Ne und Hg, Phys. Zs. **22**, 417 (1921).

Adsorption und übereinstimmende Zustände (with R. Lorenz), Z. Anorg. Allgem. Chem. **125**, 47 (1922).

Zur Theorie der anomalen Zeeman- und magneto-mechanischen Effekte, Z. Physik **11**, 353 (1922).

Termstruktur und Zeemaneffekt der Multiplets, Z. Physik **15**, 189 (1923); **19**, 112 (1923).

Fortschritte beim Zeemaneffekt, Ergeb. Exakt. Naturw. **2**, 147 (1923).

Zur Theorie der Röntgenspektren, Z. Physik **16**, 391 (1923).

Zur Struktur des Neonspektrums, Z. Physik **17**, 292 (1923).

Schwierigkeiten in der Quantentheorie des Atombaus, besonders magnetischer Art, Phys. Zs. **24**, 441 (1923).

Versagen der Quantentheorie in der Mechanik, Naturwiss. **11**, 725 (1923).

Das Wesen der relativistischen Röntgendubletts, Z. Physik **24**, 88 (1924).

Die absoluten Intervalle der optischen Dubletts und Tripletts, Z. Physik **25**, 46 (1924).

Termstruktur der Multipletts höherer Stufe, Z. Physik **25**, 279 (1924).

Über gestrichene und verschobene Spektralterme, Z. Physik **27**, 149 (1924).

Über den quadratischen Zeemaneffekt, Z. Physik **30**, 329 (1924).

Lichtquanten und Kohärenz, Z. Physik **33**, 571 (1925).

Zeemaneffekt bei Multipletts höherer Stufe, Ann. Physik **76**, 273 (1925).

Zeemaneffekt und Multiplettstruktur der Spektrallinien (with E. Back) (Springer, Berlin, 1925).

Zur Quantentheorie der Strahlung, Z. Physik **35**, 317 (1926).

Ein Experiment zur Kohärenzfähigkeit von Licht (with W. Gerlach), Z. Physik **36**, 169 (1926).

Neue Wege der Quantentheorie, Naturwiss. **14**, 455 (1926).

Die Neuere Entwicklung der Quantentheorie (Steinkopf, Leipzig, 1926).

Axiomatische Begrundung der Thermodynamik durch Carathéodory, in *Handbuch der Physik* (Springer, Berlin, 1926), vol. 9, p. 281.

Spontane Quantenübergänge, Z. Physik **42**, 835 (1927).

Wellenmechanik der Kontinua und Elektrodynamik, Phys. Zs. **44**, 768 (1927).

Zu Dirac's Theorie des Kreiselelektrons, Z. Physik **48**, 601 (1928).

Optik und Thermodynamik, in *Handbuch der Physik* (Springer, Berlin, 1928), vol. 20, p. 453.

Zeemaneffekt, in *Handbuch der Physik* (Springer, Berlin, 1928), vol. 21, p. 360.

Polarisation der Materiewellen, Naturwiss. **17**, 634 (1929).

Zur Quantenelektrik von G. Mie, Z. Physik **57**, 713 (1929).

Vorlesungen über Wellenmechanik (Akademie Verlag, Leipzig, 1930).

Zur Quantenmechanik der Gasentartung, Z. Physik **74**, 780 (1933).

The Magnetic Moment of the Proton, Phys. Rev. **44**, 1028 (1933).

Neutrons in the Nucleus, Phys. Rev. **43**, 620 (1933); **43**, 624 (1933).

Nuclear Magnetic Moments and Their Origin, Phys. Rev. **46**, 477 (1934).

Waves and Corpuscles in Quantum Physics, Science **85**, 210 (1937).

Principles of Quantum Mechanics (Cambridge University Press, London, 1937).

Transitions between Levels Spaced Almost Continuously, Phys. Rev. **54**, 940 (1938).

Critical Remarks on the Interpretation of Quantum Theory, J. Franklin Inst. **226**, 83 (1938).

Sommerfeld's Fine Structure Constant and Born's Reciprocity, J. Franklin Inst. **228**, 495 (1939).

On the Existence and the Magnitude of Electronic Charges, J. Franklin Inst. **229**, 767 (1940).

On the Stability and Magnitude of Electronic Charges (with L. H. Thomas), J. Franklin Inst. **231**, 63 (1941).

On the Magnitude of Electronic Charges, Phys. Rev. **59**, 434 (1941).

Finite Self-energies in Radiation Theory (with L. H. Thomas), Phys. Rev. **60**, 121 (1941); **60**, 541 (1941); **65**, 175 (1944).

Interaction between Elementary Particles, Phys. Rev. **76**, 1176 (1949); **77**, 814 (1950).

On Advanced and Retarded Potentials, Phys. Rev. **80**, 283 (1950).

Quantum Mechanics (Pitman, London, 1950).

Thermodynamic Continuity and Quantum Principles, Phys. Rev. **87**, 267 (1952).

Quantum Mechanics and Thermodynamic Continuity, Am. J. Phys. **20**, 353 (1952); **22**, 82 (1954).

Continuity, a Key to Quantum Mechanics, J. Phil. Sci. **20**, 101 (1953).

Quantum Mechanics, a Thermodynamic Approach, Am. Scientist **41**, 439 (1953).

Probability in Classical and Quantum Theory, in *Max Born Anniversary Volume* (Oliver and Boyd, Edinburgh, 1953), p. 59.

Thermodynamische Begründung der Quantenmechanik, Naturwiss. **41**, 125 (1954); **41**, 524 (1954).

Quantum Indeterminacy, a Consequence of Cause-Effect Continuity, Dialectica **8**, 199 (1954).

Le Principe de Continuité et la Théorie des Quanta, J. Phys. Radium **16**, 353 (1955).

Foundations of Quantum Theory (Yale University Press, New Haven, 1955).

Quantum Mechanics and Common Sense, Endeavour, vol. 15, 1956.

ψ-Superposition and Quantum Rules, Am. J. Phys. **24**, 56 (1956).

The Logic of Quanta, Brit. J. Phil. Sci. **6**, 300 (1956).

Déduction de la Théorie quantique à Partir de Principes non-Quantiques, J. Phys. Radium **17**, 1 (1956).

Quantentheorie auf nicht-quantenhafter Grundlage, Naturwiss. **10**, 217 (1956).

ψ-Superposition and Quantum Periodicity, Phys. Rev. **108**, 891 (1957).

Wellenmechanik und Irreversibilität, Physik. Bl. **13**, 312 (1957).

Non-Quantal Foundations of Quantum Theory, J. Phil. Sci. **24**, 309 (1957).

Quantum Physics and Philosophy, Current Sci. **27**, 81 (1958).

Quantum Theory from Non-Quantal Postulates, in *Berkeley Symposium on the Axiomatic Methods* (1958), p. 353.

Determinism versus Continuity, Mind **67**, 266 (1958).

Ist die Dualität in der Quantentheorie ein Erkenntnisproblem? Philosophia Naturalis **5**, 498 (1958).

Zur Quantentheorie der Messung, Z. Physik **153**. 389 (1959).

Can Physical Knowledge Be Satisfied with a Dualistic Picture Rather than a Unitary Reality? Marquette University Symposium, June 1959.

Heisenberg's Contracting Wave Packets, Am. J. Phys. **27**, 415 (1959).

Quantum Mechanics from Duality to Unity, Am. Scientist **47**, 341 (1959).

From Dualism to Unity in Quantum Mechanics, Brit. J. Phil. Sci. **10**, 16 (1959).

From Dualism to Unity in Quantum Theory (Cambridge University Press, London, 1960).

From Duality to Unity in Quantum Mechanics, in *Current Issues in the Philosophy of Science*, ed. by H. Feigl and G. Maxwell (Holt, Rinehart and Winston, New York, 1961).

Unitary Interpretation of Quantum Theory, Am. J. Phys. **29**, 503 (1961).

Dualismus, Wissenschaft und Hypothese, in *Heisenberg Festschrift* (Vieweg, Brunswick, 1961), p. 119.

Ableitung der Quantenregeln auf nicht-quantenmässiger Grundlage, Z. Physik **162**, 410 (1961).

Warum Interferieren die Wahrscheinlichkeiten? Z. Physik **164**, 558 (1961).

The Case against Quantum Duality, J. Phil. Sci. **29**, 1 (1962).

Why Do Quantum Theorists Ignore the Quantum Theory? Brit. J. Phil. Sci. **15**, 307 (1963).

Von Dualismus zur Einheitlichen Quantentheorie, Phil. Nat. **8**, 232 (1964).

Solution of the Gibbs Paradox, J. Phil. Sci. **32**, 192 (1965).

Non-Quantal Foundations of Quantum Mechanics, Dialectica **19**, 349 (1965).

Quantum Fact and Fiction, Am. J. Phys. **33**, 123 (1965); **34**, 1160 (1966); **37**, 541 (1969).

New Foundations of Quantum Mechanics (Cambridge University Press, London, 1965).

Quantum Theory without Dualism, Scientia **7**, 1 (1966).

Observation and Interpretation in Quantum Theory, in *Proceedings of the Seventh Inter-American Congress of Philosophy*, Laval University, 1967.

New Foundations of Quantum Physics, Physics Today **20**, 55 (1967).

Quantum Physics and Philosophy, in *Contemporary Philosophy*, ed. by R. Klibansky (La Nuova Italia Editrice, 1968).

Quantum Observation and Interpretation, in *XIV. Internationaler Kongress für Philosophie* (Herder, Vienna, 1968).

Quantenmechanik, Beobachtung und Deutung, Intern. J. Theoret. Phys. **1**, 51 (1968).

Dualismus in der Quantentheorie, Eine Entgegnung, Philosophia Naturalis **11**, 395 (1969).

Wahrheit und Dichtung in der Quantentheorie, Physik. Bl. **25**, 105 (1969).

Non-Quantal Foundations of Quantum Mechanics, in *Physics, Logic, and History*, ed. by W. Yourgrau and A. Breck (Plenum Press, New York and London, 1970).

List of Contributors

W. Yourgrau
Foundations of Physics
University of Denver
U.S.A.

A. J. van der Merwe
Department of Physics
University of Denver
U.S.A.

M. Born
(1882–1970)

L. de Broglie
Faculté des Sciences
Physique Théorique
Institut Henri Poincaré
France

H. -J. Treder
Kosmische Physik
Deutsche Akademie
der Wissenschaften
German Democratic Republic

E. P. Wigner
Department of Physics
Joseph Henry Laboratories
Princeton University
U.S.A.

J. L. Park
Department of Physics
Washington State University
U.S.A.

H. Margenau
Sloane Physics Laboratory
Yale University
U.S.A.

A. Kastler
École Normale Supérieure
Laboratoire de Physique
Université de Paris
France

D. Bohm
Birkbeck College
Department of Physics
University of London
Britain

H. Hönl
Institut für theoretische Physik
Universität Freiburg
German Federal Republic

D. Caldwell
Department of Chemistry
University of Utah
U.S.A.

H. Eyring
Department of Chemistry
University of Utah
U.S.A.

O. Costa de Beauregard
Faculté de Sciences
Physique Théorique
Institut Henri Poincaré
France

F. Bopp
Theoretische Physik
Universität München
German Federal Republic

J. -P. Vigier
Faculté de Sciences
Physique Théorique
Institut Henri Poincaré
France

K. R. Popper
Department of Logic
and Scientific Method
London School of Economics
and Political Science
Britain

W. M. Elsasser
Institute for Fluid Dynamics
and Applied Mathematics
University of Maryland
U.S.A.

L. Rosenfeld
Nordita
Denmark

H. Bondi
European Space Research
Organisation
92 Neuilly-sur-Seine
France

A. Mercier
Institut für theoretische Physik
Universität Bern
Switzerland

P. Bernays
8002 Zürich
Bodmerstrasse 11
Switzerland

Name Index

Subject Index